How Light Transmits Sound

By: Dr. Joseph Mucha, J.D.
       AI Prompt Doctor

Table of Contents

Quantum Optomechanics: Harnessing Quantum
Effects for Sound Transduction

X.   Conclusion

XI.   Appendix

I.   Introduction

In the vast realm of scientific discoveries and phenomena, few are as captivating and intriguing as the interaction between light and sound. While conventionally viewed as distinct entities, recent advancements have revealed a profound connection between these two fundamental aspects of our world. "How Light Transmits Sound" delves into this captivating relationship, unearthing the mesmerizing interplay between light waves and sound waves, and the profound implications it holds for our understanding of the universe.

Throughout history, light and sound have enthralled scientists, philosophers, and artists alike. From the ancient Greeks pondering the nature of vision and hearing to the groundbreaking experiments of luminaries like Isaac Newton and Thomas Young, humans have sought to unravel the mysteries behind these phenomena. However, it is only in recent decades that we have witnessed remarkable breakthroughs, unveiling the intricate ways in which light and sound are intertwined.

This book serves as a comprehensive guide, navigating the intricate terrain where light and sound merge, coalesce, and give rise to new possibilities. From the fundamental principles underlying the propagation of light and sound to cutting-edge research at the forefront of scientific exploration, "How Light

Transmits Sound" explores the multidimensional nature of this remarkable relationship.

Within these pages, readers will embark on a journey that traverses various disciplines, including physics, acoustics, optics, and materials science. Engagingly written, this book combines accessible explanations with captivating examples, making complex concepts relatable and tangible to both experts and enthusiasts.

Discover the intriguing phenomenon of "sonoluminescence," where sound waves create flashes of light in tiny bubbles, illuminating the hidden connections between light and sound. Delve into the fascinating world of optoacoustic imaging, where laser pulses convert absorbed light energy into ultrasound waves, enabling non-invasive exploration of biological tissues. Witness the breakthroughs in fiber-optic communications, where light signals are converted to sound and back again, revolutionizing data transmission.

Moreover, "How Light Transmits Sound" examines the practical applications of this intertwined relationship, ranging from the development of advanced medical imaging techniques to the design of novel acoustic metamaterials with unprecedented control over sound propagation. The potential implications extend even further, encompassing areas such as telecommunications, underwater acoustics, and even the study of exoplanets.

As we embark on this enlightening journey, let us unravel the mysteries of this entwined duality. Through meticulous research, thought-provoking insights, and captivating anecdotes, this book seeks to illuminate the fascinating world where light and sound converge, revolutionizing our perception of the universe and inspiring future explorations.

Join us in the exploration of "How Light Transmits Sound," where the symphony of scientific knowledge unfolds, revealing the harmonious dance of these fundamental forces.

The Fascinating Connection Between Light and Sound

In the tapestry of the natural world, there exists a captivating interplay between two fundamental phenomena: light and sound. These ethereal forces, seemingly distinct, are intricately linked in ways that have captivated the minds of scientists, artists, and philosophers throughout the ages. The fascinating connection between light and sound unravels a rich tapestry of knowledge, weaving together intricate patterns that deepen our understanding of the universe.

Light, with its radiant energy and capacity to reveal the world around us, has long been a subject of wonder. Sound, the vibrational waves that fill the air and resonate within us, stirs emotions and connects us to our surroundings. While we commonly perceive them through separate senses, they share a profound relationship that transcends mere perception.

The genesis of this connection can be traced back to the underlying physics that governs both light and sound. Both phenomena arise from the fundamental nature of waves and their propagation through space or matter. They share common properties such as wavelength, frequency, amplitude, and velocity, albeit in different ranges and contexts.

In recent years, scientific exploration has unveiled astonishing insights into the interconnectedness of light and sound. Researchers have discovered extraordinary phenomena, such as the ability of sound to generate light, and vice versa. These discoveries have shattered traditional boundaries and opened up new realms of study, fostering collaborations between disciplines that were once considered distinct.

One of the most intriguing phenomena that exemplifies the connection between light and sound is the emerging field of optoacoustics. Here, light energy is used to generate acoustic waves, providing a powerful imaging tool with applications in medicine, biology, and materials science. By harnessing laser-induced sound waves, researchers can explore the inner workings of tissues, visualize biological structures, and even detect and treat diseases at the cellular level.

Another fascinating example lies in the world of nanotechnology, where scientists manipulate light and sound at the smallest scales. Nanostructures can be engineered to interact with both light and sound waves simultaneously, leading to novel applications in sensing, energy conversion, and data storage. These advancements blur the boundaries between optics and acoustics, giving rise to a new frontier of research and technological possibilities.

Moreover, the connection between light and sound extends beyond the realm of human perception. It plays a crucial role in diverse natural phenomena, from the mesmerizing displays of color and sound in the auroras to the intricate communication systems of animals and the harmonious melodies of music. This intrinsic relationship between light and sound pervades the fabric of the natural world, providing a profound insight into the interconnectedness of all things.

As we delve deeper into the mesmerizing connection between light and sound, we embark on a journey of discovery, curiosity, and wonder. Through meticulous experimentation, theoretical exploration, and artistic expression, we uncover the hidden symphony that unites these fundamental forces. This exploration paves the way for groundbreaking advancements, inspiring innovations that impact fields as diverse as communication, entertainment, healthcare, and beyond.

In this narrative of exploration, we invite you to join us on a voyage through the captivating realms where light and sound converge. Together, we shall illuminate the profound interconnectedness of these phenomena and celebrate the harmonious bond that enriches our understanding of the universe.

Overview of Light and Sound Transmission

In the realm of energy propagation, two fundamental phenomena hold sway: light and sound. Light, the radiant energy that illuminates our world, and sound, the vibrational waves that permeate our surroundings, possess unique characteristics and pathways of transmission. Understanding how light and sound propagate is essential to unraveling their far-reaching implications across various scientific disciplines and everyday life. In this overview, we delve into the fascinating world of light and sound transmission, exploring the mechanisms that govern their journeys through space, materials, and our senses.

Light Transmission:

Light, characterized by its diverse spectrum of wavelengths, plays a pivotal role in our perception of the world. It propagates through space as electromagnetic waves, guided by the laws of optics. The transmission of light is influenced by several factors, including the nature of the medium it traverses and its interaction with objects along its path.

In a vacuum, light travels at a constant speed of approximately 299,792 kilometers per second, known as the speed of light. However, when light encounters a medium, such as air, water, or glass, its speed is altered due to interactions with the atoms and molecules of the material. This phenomenon, known as refraction, causes the bending of light rays as they pass from one medium to another, resulting in effects like the splitting of white light into its constituent colors in a prism.

Additionally, light can be absorbed, reflected, or transmitted when it encounters an object. Absorption occurs when light energy is absorbed by the atoms or molecules of the material, converting it into other forms of energy, such as heat. Reflection takes place when light bounces off a surface, following the law of reflection that states the angle of incidence is equal to the angle of reflection. Transmission occurs when light passes through a material without being significantly absorbed or reflected, allowing it to propagate beyond the object.

Understanding light transmission is vital in fields such as optics, photography, telecommunications, and even astronomy. Fiber-optic communication systems utilize the transmission of light through thin, flexible glass or plastic fibers to transmit data over long distances at high speeds. Telescopes capture and analyze the faint light from distant celestial objects, unraveling the secrets of the cosmos. In everyday life, the way light transmits through materials influences the design of transparent windows, lenses, and displays, enhancing our visual experiences.

Sound Transmission:

Sound, as a mechanical wave, propagates through various media, primarily air, liquids, and solids. It originates from a source, such as a vibrating object, and travels through these media in the form of compressions and rarefactions, creating alternating regions of higher and lower pressure.

In gaseous media, like air, sound waves propagate through molecular collisions. When an object vibrates, it compresses and rarefies the surrounding air particles, creating a chain reaction of compressions and rarefactions that travel away from the source. The speed of sound varies with the properties of the medium, such as temperature, humidity, and composition. In dry air at sea level, sound typically travels at a speed of around 343 meters per second.

When sound encounters a different medium, such as water or a solid object, its transmission is influenced by the density and elasticity of the material. In liquids and solids, sound waves propagate through the vibration and interaction of molecules or atoms. The specific properties of the material, such as its density, stiffness, and internal structure, determine how sound travels and the characteristics of its propagation, including speed, direction, and attenuation.

Understanding sound transmission is essential in fields such as acoustics, audio engineering, architecture, and communication. Architectural acoustics studies the transmission and manipulation of sound within buildings, optimizing designs to enhance speech intelligibility, minimize echoes, and create pleasant listening environments. Audio engineers manipulate sound transmission to capture, reproduce, and amplify audio signals, ensuring high-quality recordings and immersive experiences. In telecommunications, sound transmission plays a crucial role in the development of efficient and reliable communication systems, such as telephones, microphones, and speakers.

The Interplay of Light and Sound Transmission:

While light and sound transmission are distinct phenomena, they often intersect and influence each other in fascinating ways. For instance, the transmission of light can be affected by the presence of sound waves, leading to phenomena like the diffraction of light around a vibrating object. Similarly, sound waves can interact with light, resulting in phenomena such as sonoluminescence, where sound waves generate tiny bursts of light in specific conditions.

In recent years, advancements in materials science have enabled the manipulation of both light and sound transmission simultaneously. Metamaterials, engineered structures with unique properties not found in nature, have been designed to control and manipulate the propagation of both light and sound waves. These materials offer unprecedented control over the behavior of waves, opening up possibilities for innovative devices and applications, such as invisibility cloaks, acoustic lenses, and enhanced optical sensors.

Moreover, the field of optoacoustics explores the interaction between light and sound, harnessing their complementary properties for imaging and sensing applications. By using laser-induced sound waves and detecting the resulting acoustic signals, researchers can visualize and probe biological tissues with remarkable precision and depth, revolutionizing medical imaging and diagnostics.

The study of light and sound transmission is a captivating journey that unveils the intricate pathways of energy propagation. Understanding how light and sound traverse through space, materials, and our senses allows us to manipulate, harness, and enhance their properties for various applications. The interplay between light and sound transmission provides fertile ground for groundbreaking discoveries, technological advancements, and new avenues of exploration across diverse scientific disciplines.

As we continue to unravel the secrets of light and sound transmission, we unlock new opportunities to deepen our understanding of the natural world, improve our communication systems, enhance our experiences, and gain insights into the profound connections that exist within the fabric of our universe.

II.   Fundamentals of Light

Light, a mesmerizing phenomenon that surrounds us, is much more than just a source of illumination. It is a fundamental aspect of our existence, shaping our perception of the world and holding the key to vast realms of scientific discovery. Understanding the fundamentals of light is essential to unraveling its profound nature and exploring its far-reaching implications in fields ranging from physics and optics to communication and technology. In this overview, we delve into the fundamental aspects of light, exploring its nature, behavior, and the remarkable properties that make it such a captivating force.

Light as an Electromagnetic Wave:

At its core, light is a form of electromagnetic radiation—a wave composed of oscillating electric and magnetic fields that propagate through space. It belongs to a broad spectrum of electromagnetic waves, encompassing radio waves, microwaves, infrared radiation, visible light, ultraviolet radiation, X-rays, and gamma rays. These waves differ in their wavelengths and frequencies, with visible light occupying a small portion of the spectrum.

The Dual Nature of Light: Particle and Wave:

One of the most intriguing aspects of light is its dual nature, displaying characteristics of both particles and waves. This duality is encapsulated in the wave-particle duality theory, which asserts that light can exhibit wave-like behavior, such as interference and diffraction, as well as particle-like behavior, known as photons.

According to the particle theory of light proposed by Albert Einstein, light can be viewed as a stream of discrete particles, each carrying a specific amount of energy known as a photon. Photons possess properties like energy, momentum, and wavelength, allowing them to interact with matter and participate in various phenomena, such as the photoelectric effect and Compton scattering.

The Wave Behavior of Light:

While the particle nature of light explains certain phenomena, its wave behavior reveals a wealth of fascinating phenomena. When light waves encounter obstacles or pass through narrow slits, they exhibit phenomena such as diffraction and interference, similar to other types of waves.

Diffraction occurs when light waves encounter an obstacle or an aperture, causing them to bend or spread out, resulting in the bending of light around corners and the creation of patterns of light and dark regions. Interference occurs when two or more light waves superpose, leading to constructive or destructive interference, where the waves either reinforce or cancel each other out, respectively.

The Speed of Light and Refraction:

One of the fundamental properties of light is its incredible speed. In a vacuum, light travels at a constant speed of approximately 299,792 kilometers per second, often rounded to 300,000 kilometers per second for simplicity. This remarkable speed sets the ultimate cosmic speed limit, known as the speed of light in vacuum.

When light encounters a different medium, such as air, water, or glass, its speed changes due to interactions with the atoms and molecules of the material. This change in speed causes the phenomenon of refraction, where light rays bend as they pass from one medium to another. The bending of light leads to effects such as the dispersion of white light into its constituent colors when passing through a prism.

The Electromagnetic Spectrum and Visible Light:

The electromagnetic spectrum encompasses a wide range of electromagnetic waves, with visible light occupying a small portion of this spectrum. Visible light consists of a continuous spectrum of colors, ranging from red, with longer wavelengths and lower frequencies, to violet, with shorter wavelengths and higher frequencies. Each color corresponds to a different wavelength and is perceived differently by our eyes.

The understanding of visible light and its interaction with matter has paved the way for numerous applications. From the development of optical devices like lenses and mirrors to the exploration of color perception, light plays a central role in fields such as optics, photography, displays and visual arts. Our ability to manipulate and control visible light has led to innovations in lighting technology, imaging systems, and display technologies that have revolutionized various industries.

The fundamentals of light encompass its dual nature as both a particle and a wave, its remarkable speed and ability to propagate through space, and its interaction with matter. Understanding these fundamental aspects allows us to explore the nature of light, unravel its mysteries, and harness its potential in diverse fields.

From the study of the electromagnetic spectrum and the behavior of light waves to the practical applications in optics, technology, and communication, the fundamentals of light open up a world of scientific exploration and technological advancement. As we continue to deepen our understanding of light, we unlock new possibilities for innovation, discovery, and the enrichment of our perception and understanding of the universe.

Properties of Light Waves

Light waves, as a fundamental aspect of electromagnetic radiation, exhibit a rich array of properties that govern their behavior and interactions. Understanding these properties is crucial to comprehending the nature of light and harnessing its applications in various fields, from optics and imaging to telecommunications and quantum physics. Here we'll delve into the properties of light waves, providing examples and explanations that shed light on their intriguing characteristics.

Wavelength and Frequency:
One of the fundamental properties of light waves is their wavelength, which refers to the distance between consecutive peaks or troughs of the wave. Wavelength is commonly denoted by the Greek letter lambda (λ) and is measured in units such as nanometers (nm) or meters (m). Different colors of light are associated with different wavelengths, with red light having longer wavelengths and violet light having shorter wavelengths.

Frequency, on the other hand, measures the number of complete wave cycles that pass a given point in one second. It is represented by the symbol f and is measured in hertz (Hz). Frequency and wavelength are inversely proportional, meaning that as the wavelength of a light wave increases, its frequency decreases, and vice versa. This relationship is defined by the equation: speed of light = wavelength x frequency.

Example: Visible light consists of a spectrum of colors, with red light having a longer wavelength (around 700 nm) and lower frequency, while violet light has a shorter wavelength (around 400 nm) and higher frequency.

Amplitude and Intensity:
The amplitude of a light wave refers to the maximum displacement of particles in the medium through which the wave is propagating. In simpler terms, it represents the "height" or "strength" of the wave. Amplitude determines the intensity or brightness of light, with higher amplitudes corresponding to more intense illumination.

Intensity, on the other hand, quantifies the amount of light energy passing through a given area per unit time. It depends on both the amplitude of the light wave and the area over which the energy is distributed. Intensity is measured in units such as watts per square meter (W/m^2) or lumens.

Example: When a flashlight is turned on and the beam is directed towards a surface, the amplitude of the light waves determines the brightness of the spot created, while the intensity determines the amount of light energy reaching the surface.

Reflection and Refraction:

Reflection occurs when light waves encounter a surface and bounce back. The angle at which the light wave approaches the surface, known as the angle of incidence, is equal to the angle at which it reflects, known as the angle of reflection. The law of reflection states that the incident angle and reflected angle are measured relative to the surface normal (a line perpendicular to the surface).

Refraction, on the other hand, refers to the bending of light waves as they pass from one medium to another with different optical densities. This bending occurs due to a change in the speed of light. The amount of bending depends on the angle of incidence and the refractive indices of the two media involved.

Example: When light waves strike a mirror, they undergo reflection, bouncing off the surface and forming an image. In contrast, when light passes from air to water, it refracts due to the change in the speed of light, causing the light wave to change direction.

Diffraction and Interference:
Diffraction refers to the bending or spreading out of light waves as they encounter an obstacle or pass through a narrow opening. The degree of diffraction depends on the size of the obstacle or opening relative to the wavelength of light. Larger obstacles or smaller openings produce more significant diffraction effects.

Interference occurs when two or more light waves superpose, leading to either constructive or destructive interference. Constructive interference occurs when the peaks of two or more waves align, resulting in an amplified wave. Destructive interference, on the other hand, happens when the peaks of one wave align with the troughs of another, leading to a diminished or canceled-out wave.

Example: When light passes through a narrow slit, it undergoes diffraction, causing it to spread out and create a pattern of light and dark regions on a screen behind the slit. Additionally, in interference experiments such as the double-slit experiment, constructive interference produces bright fringes, while destructive interference leads to dark fringes.

Polarization:
Polarization refers to the orientation of the electric field vector of a light wave. When light waves are polarized, the electric field oscillates in a specific direction perpendicular to the direction of wave propagation. Polarization can be linear, where the electric field oscillates in a single plane, or circular/elliptical, where the electric field follows a curved path.

Polarization has significant implications in areas such as optical filters, 3D movie technology, and communication systems, where the manipulation and control of polarized light waves are essential.

Example: Sunglasses with polarized lenses are designed to block horizontally polarized light, reducing glare from reflected surfaces such as water or car windows.

The properties of light waves reveal the intricacies of this electromagnetic phenomenon. Wavelength and frequency define the color and energy of light, while amplitude and intensity determine its brightness. Reflection, refraction, diffraction, and interference govern how light interacts with surfaces and objects. Polarization allows for the manipulation and control of light waves.

By understanding these properties, scientists, engineers, and researchers can harness the power of light waves in various applications, including imaging, telecommunications, optics, and quantum technologies. The exploration of these properties continues to unlock new possibilities, deepening our understanding of light and its role in the world around us.

The Electromagnetic Spectrum

The electromagnetic spectrum encompasses a vast range of electromagnetic waves, each characterized by its unique wavelength, frequency, and energy. This spectrum extends far beyond the visible light that our eyes perceive, spanning from low-frequency radio waves to high-energy gamma rays. Understanding the electromagnetic spectrum is essential for comprehending the diverse forms of electromagnetic radiation and their wide-ranging applications in fields such as communication, medicine, astronomy, and beyond. Here, we explore the electromagnetic spectrum, its divisions, and the fascinating phenomena it encompasses.

Radio Waves:
At the lower end of the electromagnetic spectrum lie radio waves, with the longest wavelengths and lowest frequencies. These waves are utilized in a multitude of practical applications, including wireless communication, broadcasting, and radar systems. They can easily penetrate buildings and travel long distances, making them ideal for transmitting information over vast areas.

Example: When you tune in to your favorite radio station, the broadcast signals are carried by radio waves that propagate through the air, allowing you to listen to music or news.

Microwaves:
Slightly higher in frequency and shorter in wavelength than radio waves, microwaves find applications in cooking, communication, and remote sensing. Microwave ovens use these waves to heat food by exciting water molecules. They are also employed in satellite communication, Wi-Fi networks, and weather radar systems.

Example: When you use a microwave oven to heat your meal, the microwaves generated by the oven interact with the water molecules in the food, causing them to vibrate and generate heat.

Infrared Radiation:
Beyond microwaves, we encounter infrared radiation, often referred to as heat radiation. Infrared waves have longer wavelengths than visible light and are known for their ability to transfer heat energy. They find applications in thermal imaging, remote controls, and heating systems.

Example: Infrared cameras detect and visualize the heat emitted by objects, allowing firefighters to locate individuals in smoke-filled environments or enabling scientists to study thermal patterns in various materials.

Visible Light:
Visible light is the narrow band of the electromagnetic spectrum that our eyes can detect. It comprises different colors, each corresponding to a specific wavelength

within the range of approximately 400 to 700 nanometers. Our perception of color is a result of the interaction between light waves and the receptors in our eyes.

Example: The colors we see in a rainbow, a prism experiment, or a vibrant painting are all manifestations of visible light. Each color corresponds to a specific range of wavelengths within the visible spectrum.

Ultraviolet (UV) Radiation:
Moving beyond the visible spectrum, we encounter ultraviolet radiation, which has shorter wavelengths and higher frequencies than visible light. UV radiation has both beneficial and harmful effects. It is responsible for vitamin D synthesis in our bodies and is used in sterilization processes and forensic analysis. However, overexposure to UV radiation can cause sunburn, skin damage, and an increased risk of skin cancer.

Example: Sunscreen lotions provide protection by absorbing or reflecting UV radiation, thus shielding our skin from its harmful effects.

X-Rays:
X-rays have even shorter wavelengths and higher energies than UV radiation. These waves are well-known for their ability to penetrate matter and are extensively used in medical imaging to visualize internal structures, such as bones and organs. They are also

employed in industrial inspections and security screening.

Example: When you undergo a dental or medical X-ray, X-ray waves pass through your body, and the resulting image provides valuable diagnostic information to healthcare professionals.

Gamma Rays:
At the highest end of the electromagnetic spectrum are gamma rays. These waves have the shortest wavelengths and highest energies. Gamma rays are produced by nuclear reactions and radioactive decay processes. They find applications in medical treatments, cancer therapy, and sterilization procedures. Gamma rays are also studied in astrophysics to understand high-energy phenomena in the universe.

Example: In cancer treatment, gamma rays are used in a process called radiation therapy, where high-energy beams are directed at tumors to destroy cancer cells and shrink tumors.

The electromagnetic spectrum encompasses a vast range of electromagnetic waves, each with its own unique properties and applications. From radio waves for communication to gamma rays for cancer treatment, the spectrum offers a wealth of possibilities for scientific exploration and technological advancement.

Understanding the divisions of the electromagnetic spectrum allows us to harness the power of electromagnetic radiation in various fields, from everyday communication to cutting-edge medical imaging and treatment. As technology advances and our understanding deepens, the electromagnetic spectrum continues to reveal new realms of knowledge and innovation, shaping the way we perceive and interact with the world around us.

Interaction of Light with Matter

When light encounters matter, a fascinating interplay unfolds. Light waves interact with the atoms and molecules of materials, giving rise to a wide array of phenomena that shape our perception and understanding of the world. Understanding the interaction of light with matter is crucial in fields such as optics, spectroscopy, photonics, and material science. In here we explore the captivating ways in which light and matter interact, unveiling the mechanisms that govern energy transfer and the remarkable applications that arise from these interactions.

Absorption:
Absorption occurs when light waves transfer their energy to matter, causing the atoms or molecules within the material to undergo an excitation. This process typically involves the promotion of electrons to higher energy levels. The energy of the absorbed light is converted into internal energy, often leading to temperature increase or chemical reactions.

Example: When white light strikes a red apple, the apple appears red because it absorbs most of the colors in the incident light spectrum except for red, which is reflected or transmitted.

Reflection:
Reflection is the phenomenon in which light waves encounter a surface and bounce back, returning to the

medium from which they originated. The angle of incidence, the angle at which light strikes the surface, is equal to the angle of reflection, the angle at which light reflects off the surface. The reflective properties of materials depend on factors such as their composition, texture, and surface characteristics.

Example: When sunlight hits a mirror, the mirror's smooth and polished surface reflects the light, enabling us to see our reflection or to redirect light for various practical purposes.

Refraction:
Refraction occurs when light waves pass from one medium to another with a different optical density, causing a change in direction and speed. This change is due to the variation in the light's velocity as it transitions between media. The bending of light at the interface of the two media is governed by Snell's law, which relates the angles of incidence and refraction to the refractive indices of the two materials involved.

Example: When light waves pass from air into water, they slow down and bend, causing objects submerged in water to appear shifted or distorted relative to their actual position.

Scattering:
Scattering is the process in which light waves interact with particles or irregularities in a material, causing the light to deviate from its original path. Scattering can

occur in different ways, such as Rayleigh scattering, which is responsible for the blue color of the sky, or Mie scattering, which produces the white color of clouds. The scattering of light can result in phenomena like diffusion, haze, and the formation of colorful optical effects.

Example: When sunlight passes through a cloudy sky, the tiny water droplets or ice crystals scatter the light, giving rise to the diffuse, soft lighting conditions that are often associated with cloudy weather.

Transmission:
Transmission refers to the propagation of light waves through a material, without significant absorption or reflection. Transparent materials allow light to pass through them with minimal attenuation, whereas translucent materials scatter and partially transmit light. The transmission properties of materials depend on their molecular structure and the energy levels of their constituent atoms or molecules.

Example: When light passes through a pane of glass or a clear plastic sheet, it transmits through the material, allowing us to see objects on the other side.

Emission:
Emission occurs when matter releases energy in the form of light waves. This process can happen through various mechanisms, such as thermal emission, fluorescence, phosphorescence, or stimulated emission

(as in lasers). Each mechanism involves the release of energy from excited states of atoms or molecules, resulting in the emission of light with specific characteristics.

Example: Fluorescent lamps emit visible light by absorbing energy from an external source, such as an electric current, exciting the atoms or molecules within the lamp. As these excited states relax, they emit visible light, creating illumination.

The interaction of light with matter is a captivating interplay that shapes our perception of the world and enables a multitude of technological applications. From absorption and reflection to refraction, scattering, transmission, and emission, the behavior of light waves when encountering matter unveils a rich tapestry of phenomena.

Understanding these interactions allows us to manipulate and harness light for various purposes, ranging from the design of optical devices and materials to the development of advanced imaging techniques, communication systems, and light-based technologies. As we deepen our understanding of the intricate dance between light and matter, we unlock new realms of scientific exploration and technological advancements, further expanding our understanding of the universe and enhancing our capabilities in numerous fields.

III.   Basics of Sound Waves

Sound waves are an integral part of our daily lives, allowing us to communicate, experience music, and perceive the world around us. Understanding the basics of sound waves is essential for comprehending the physics of sound, the mechanisms of hearing, and the applications in fields such as acoustics, music, and communication. In this overview, we delve into the fundamental aspects of sound waves, exploring their characteristics, propagation, and the remarkable phenomena that arise from their vibrations.

Nature of Sound Waves:
Sound waves are mechanical waves that result from the vibrations of particles in a medium. These waves transfer energy from a sound source to our ears, enabling us to perceive sound. Unlike electromagnetic waves, which can travel through a vacuum, sound waves require a medium, such as air, water, or solids, to propagate.

Waveform and Frequency:
Sound waves exhibit a characteristic waveform, which represents the variation of pressure or displacement of particles in the medium over time. The waveform of a sound wave determines its unique characteristics, such as its timbre, pitch, and quality.

Frequency is a fundamental property of sound waves, referring to the number of complete cycles or vibrations per second. It is measured in hertz (Hz). Higher frequencies result in higher-pitched sounds, while lower frequencies produce lower-pitched sounds.

Example: A high-pitched whistle produces a sound wave with a high frequency, while a low-pitched drum produces a sound wave with a low frequency.

Amplitude and Intensity:
Amplitude represents the maximum displacement or pressure variation of particles in a sound wave. It determines the loudness or volume of the sound. The greater the amplitude, the louder the sound.

Intensity, on the other hand, measures the amount of energy carried by the sound wave per unit area. It depends on both the amplitude of the wave and the distance from the source. Intensity is measured in units such as watts per square meter (W/m^2).

Example: When you increase the volume of a speaker, you are effectively increasing the amplitude of the sound waves, resulting in a louder sound.

Propagation of Sound Waves:
Sound waves propagate through a medium by creating areas of compression (increased pressure) and rarefaction (decreased pressure). The particles in the medium oscillate back and forth, transferring the sound energy from one particle to the next.

The speed of sound depends on the properties of the medium, such as its density, temperature, and elasticity. In dry air at sea level and room temperature, sound generally travels at approximately 343 meters per second.

Example: When you speak, the sound waves generated by your vocal cords travel through the air, reaching the ears of the listener and enabling them to hear your voice.

Reflection, Refraction, and Diffraction:
Similar to light waves, sound waves can undergo reflection, refraction, and diffraction.

Reflection occurs when sound waves encounter a surface and bounce back. For instance, when sound waves strike a wall, they reflect off the surface, allowing us to hear echoes.

Refraction occurs when sound waves change direction as they pass from one medium to another with different properties. This change in direction is influenced by factors such as temperature gradients in the medium.

Diffraction refers to the bending or spreading out of sound waves when they encounter obstacles or pass through openings. It is particularly significant when the size of the obstacle or opening is comparable to the wavelength of the sound wave.

Example: In a concert hall, sound waves reflect off the walls, ceiling, and other surfaces, creating a rich and immersive listening experience.

Interference and Resonance:
Interference occurs when two or more sound waves overlap, resulting in constructive or destructive interference. Constructive interference amplifies the sound wave, creating a louder and stronger signal, while destructive interference leads to a reduction or cancellation of the sound wave.

Resonance is a phenomenon that occurs when an object or medium vibrates at its natural frequency in response to an external sound wave. This amplifies the sound and can lead to enhanced vibrations and prolonged sound production.

Example: When two musical instruments play the same note simultaneously, the sound waves they produce can interfere constructively, resulting in a louder and richer sound. Alternatively, if two sound waves are out of phase and interfere destructively, they can cancel each other out, leading to a reduction in sound.

The basics of sound waves reveal the intricate nature of acoustic phenomena. Waveform, frequency, amplitude, and intensity shape our perception of sound. Understanding the propagation of sound waves through various media and their interactions with surfaces and objects allows us to engineer better acoustic environments, design efficient communication systems, and create immersive musical experiences.

As we deepen our understanding of sound waves, we unlock new possibilities in fields such as acoustics, audio engineering, medicine, and telecommunications. Exploring the characteristics and behaviors of sound waves continues to expand our knowledge of the audible world and enriches our appreciation of the symphony of vibrations that surrounds us.

Characteristics of Sound Waves

Sound waves, with their ability to convey information, evoke emotions, and shape our perception of the world, hold a prominent place in our lives. Understanding the characteristics of sound waves is essential for comprehending their properties, behaviors, and applications in fields such as acoustics, music, communication, and medicine. In this overview, we delve into the fundamental characteristics of sound waves, exploring their attributes, interactions, and the remarkable phenomena they manifest.

Frequency and Pitch:
Frequency is a fundamental characteristic of sound waves, representing the number of cycles or vibrations the wave completes per unit of time. It is measured in hertz (Hz), where 1 Hz corresponds to one cycle per second. Frequency determines the perceived pitch of a sound. Higher-frequency sound waves produce higher-pitched sounds, while lower-frequency waves create lower-pitched sounds.

Example: A high-pitched whistle produces a sound wave with a high frequency, such as 10,000 Hz, resulting in a shrill sound. In contrast, a low-pitched drumbeat may have a frequency of 100 Hz, producing a deep and resonant sound.

Amplitude and Loudness:

Amplitude refers to the maximum displacement or variation in pressure of particles in a sound wave. It represents the intensity or strength of the wave and determines the perceived loudness or volume of the sound. Greater amplitude corresponds to a louder sound, while smaller amplitude results in a softer sound.

Example: When you increase the volume on a speaker, you are effectively increasing the amplitude of the sound waves, resulting in a louder and more audible sound.

Wavelength and Spatial Characteristics:

Wavelength represents the distance between two consecutive points of a sound wave that are in phase. It is typically measured from peak to peak or trough to trough. Wavelength is inversely related to frequency, meaning that shorter wavelengths correspond to higher frequencies, and longer wavelengths correspond to lower frequencies.

Spatial characteristics, such as the distance between sound sources and listeners, influence the perception of sound. The spreading of sound waves over distance can lead to a decrease in intensity and changes in the balance of frequencies.

Example: In a musical performance, higher-pitched instruments, such as flutes, produce sound waves with shorter wavelengths, while lower-pitched instruments, such as tubas, generate sound waves with longer wavelengths.

Timbre and Harmonics:
Timbre refers to the unique quality or character of a sound, which allows us to distinguish different instruments or voices even when they play the same pitch at the same loudness. It is influenced by the presence and distribution of harmonics in a sound wave.

Harmonics are additional frequencies present in a sound wave that are integer multiples of the fundamental frequency. They contribute to the complex waveform and tonal richness of a sound.

Example: A piano and a guitar playing the same note at the same loudness will have different timbres due to the distinct harmonic structure of their sound waves.

Phase and Interference:
Phase refers to the relative position or alignment of two or more sound waves at a given point in time. It influences the nature of interference when multiple sound waves interact.

Interference occurs when two or more sound waves combine at a specific point, leading to constructive or destructive interference. Constructive interference amplifies the sound, while destructive interference results in the cancellation or reduction of the sound.

Example: When two musical instruments play the same note simultaneously, the sound waves they produce can interfere constructively, resulting in a richer and louder sound. Alternatively, if the waves interfere destructively, they can cancel each other out, leading to a softer sound or even silence.

The characteristics of sound waves offer a fascinating insight into the symphony of vibrations that permeate our acoustic world. Frequency determines the pitch, amplitude influences the loudness, and wavelength provides spatial information. Timbre adds a unique color to sounds, while harmonics contribute to their richness. Phase and interference dictate the nature of wave interactions, shaping the quality and intensity of sound.

Understanding these characteristics enables us to manipulate and control sound waves in various applications. From designing musical instruments and audio systems to optimizing room acoustics and developing communication technologies, the knowledge of sound wave characteristics allows us to create immersive experiences, enhance communication, and improve our understanding of the auditory environment.

As we continue to explore the intricacies of sound waves, we unravel new possibilities in fields such as acoustics, music production, speech recognition, and medical diagnostics. The study of these characteristics expands our perception of sound and deepens our appreciation of the diverse tapestry of vibrations that form the sonic landscape around us.

Sound Propagation through Different Media

Sound waves, with their ability to traverse different mediums, allow us to communicate, experience music, and perceive the world around us. Understanding how sound propagates through various media is essential for comprehending its behavior, characteristics, and applications in fields such as acoustics, communication, and underwater exploration. This overview, we delve into the propagation of sound waves through different mediums, exploring the unique phenomena and challenges that arise in each scenario.

Air:
Air is the most common medium through which sound propagates in our everyday environment. When a sound source, such as a speaker or a musical instrument, generates vibrations, it causes the adjacent air particles to oscillate back and forth, transferring the sound energy. The sound waves propagate as a series of compressions and rarefactions, creating areas of increased and decreased air pressure.

Example: When you speak, the sound waves generated by your vocal cords travel through the air, reaching the ears of listeners and allowing them to hear your voice.

Water:
Water is denser and less compressible than air, which affects the propagation of sound waves through this medium. In water, sound waves travel at a higher speed

compared to air due to the increased density. The molecular interactions and the close proximity of water molecules allow sound waves to propagate more efficiently, leading to less loss of energy during transmission.

Example: In marine environments, animals like dolphins and whales use sound waves to communicate and navigate through the water, as their hearing and vocalization abilities are adapted to underwater sound propagation.

Solids:
In solids, such as metals or building materials, sound waves propagate differently compared to air or water. Solids are characterized by tightly packed particles, allowing sound waves to travel as vibrations through the atomic or molecular lattice. This property gives solids a high speed of sound propagation and the ability to transmit sound over long distances without significant loss of energy.

Example: When you strike a metal bell, the vibrations created by the impact travel through the solid metal structure, producing a characteristic ringing sound.

Vacuum:
Unlike air, water, or solids, a vacuum is a complete absence of matter. Therefore, sound waves cannot propagate through a vacuum. In the absence of

molecules or particles to transfer the sound energy, there is no medium for sound waves to travel.

Example: In outer space, where the vacuum prevails, sound cannot propagate as it does on Earth. This is why we often refer to space as being silent.

Unique Media:
Certain specialized media exhibit distinct sound propagation characteristics due to their specific properties. For example:

Underwater environments: Sound waves in underwater environments can exhibit long-distance propagation and low attenuation. However, sound speed and direction can be influenced by factors such as temperature, salinity, and depth variations in the water.

Gases other than air: Sound propagation through different gases, such as helium or sulfur hexafluoride, can lead to altered characteristics due to variations in density and molecular interactions.

Biological tissues: Sound waves can propagate through biological tissues, enabling medical ultrasound imaging and therapy. Different tissues have varying acoustic properties, affecting the speed and attenuation of sound waves.

Example: Ultrasound imaging uses sound waves to visualize internal structures in the human body. The waves propagate through tissues, reflecting off boundaries and generating images that aid in medical diagnosis.

The propagation of sound waves through different mediums showcases the adaptability and versatility of this acoustic phenomenon. Whether traveling through air, water, solids, or unique media, sound waves offer insights into our environment and provide a means of communication and exploration.

Understanding the propagation characteristics of sound waves allows us to optimize communication systems, design acoustic spaces, develop underwater technologies, and advance medical diagnostics. The study of sound propagation through different media enhances our ability to harness sound's potential, furthering our understanding of the acoustic world in which we live.

Perception of Sound

Sound is not merely a physical phenomenon but a rich sensory experience that shapes our perception of the world. The perception of sound involves the intricate workings of our auditory system, from the mechanics of the ear to the neural processing in the brain. Understanding the perception of sound allows us to appreciate the complexities of auditory sensation, the nuances of music, and the communication power of speech. In this overview, we delve into the fascinating aspects of sound perception, exploring the mechanisms that underlie our auditory experiences.

Sound Reception: The Ear and Auditory System
The perception of sound begins with the reception of sound waves by the ear. The outer ear collects sound waves and channels them into the ear canal. The waves then reach the middle ear, where they cause vibrations of the eardrum and tiny bones called ossicles. These vibrations are transferred to the fluid-filled cochlea, the sensory organ of the inner ear.

Within the cochlea, specialized hair cells convert the mechanical vibrations into electrical signals. These signals are then transmitted via the auditory nerve to the brain for further processing and interpretation.

Example: When you hear a piece of music, the sound waves travel through your outer ear, causing vibrations of the eardrum and the ossicles. These vibrations are then converted into electrical signals by the hair cells in the cochlea, enabling you to perceive and enjoy the music.

Pitch Perception:
Pitch refers to the perceived frequency of a sound wave. It determines whether a sound is perceived as high or low. The human auditory system can detect a wide range of frequencies, typically from 20 Hz to 20,000 Hz. Different regions within the cochlea are specialized for detecting specific frequency ranges.

Example: When you hear a high-pitched violin note, it is because the sound wave has a high frequency, while a low-pitched bass note corresponds to a sound wave with a low frequency.

Loudness Perception:
Loudness is the perception of the intensity or amplitude of a sound wave. It relates to how soft or loud a sound is perceived. The amplitude of the sound wave corresponds to the physical energy of the sound.

The perception of loudness is not solely determined by the physical amplitude but is influenced by the sensitivity of our auditory system. The loudness perception follows a logarithmic scale, meaning that a doubling of the physical sound intensity does not result in a perceived doubling of loudness.

Example: When you listen to a symphony orchestra, the brass instruments produce louder sounds compared to the string instruments due to the greater amplitude of the sound waves they generate.

Timbre Perception:
Timbre refers to the quality·or character of a sound. It allows us to distinguish between different musical instruments or recognize individual voices. Timbre is influenced by the presence and distribution of harmonics and overtones within a sound wave. These additional frequencies give each sound its unique tonal color.

Example: When you hear a piano and a guitar playing the same note, you can differentiate between the two instruments because of the distinct timbre of each sound. The piano produces a sound wave with a different harmonic structure compared to the guitar.

Spatial Perception:
Spatial perception relates to the localization and perception of sound sources in space. It allows us to determine the direction from which sound is coming. Our auditory system uses various cues, such as differences in sound arrival time, sound intensity, and spectral content, to localize sounds accurately.

Example: When you close your eyes and hear a sound coming from your left side, your auditory system utilizes the time delay between when the sound reaches your left and right ears to determine the sound's direction.

Speech Perception:
Speech perception is a specialized aspect of sound perception that enables us to understand and interpret spoken language. It involves the recognition of phonemes, which are the smallest units of sound that carry meaning in a particular language. Our auditory system processes the acoustic features of speech, such as the duration, pitch, and spectral content of phonemes, to extract meaningful information and comprehend the spoken words.

Example: When you listen to someone speaking, your auditory system analyzes the specific patterns of sound waves corresponding to different phonemes, allowing you to recognize and understand the words being spoken.

The perception of sound is a marvel of sensory processing, encompassing the reception, interpretation, and appreciation of auditory stimuli. From the intricate mechanics of the ear to the neural processing in the brain, our auditory system allows us to experience the rich tapestry of sounds that shape our world.

Understanding the mechanisms of sound perception enhances our ability to design effective communication systems, create immersive musical experiences, and diagnose auditory disorders. The perception of sound is a testament to the remarkable capabilities of the human auditory system, offering a gateway to the world of music, language, and auditory sensation.

IV.   Light-Sound Interaction:
Principles and Mechanisms

The interaction between light and sound is a fascinating area of study that reveals the interconnectedness of our senses and the complex nature of energy transfer. Understanding the principles and mechanisms of light-sound interaction allows us to unravel intriguing phenomena such as color perception, photoacoustic imaging, and optical sound modulation. In this overview, we delve into the intricate interplay between light and sound, exploring the underlying principles and mechanisms that govern this intriguing cross-sensory interaction.

Photoacoustic Effect:

The photoacoustic effect is a phenomenon that occurs when a material absorbs light energy and converts it into acoustic (sound) waves. It relies on the process of light absorption by a material, which generates local temperature variations and induces rapid thermal expansion. The resulting expansion creates pressure waves that propagate as sound waves.

Example: In photoacoustic imaging, laser light is used to illuminate tissues. When the light is absorbed by chromophores within the tissues, such as blood vessels or pigments, it generates acoustic waves that can be detected and used to create high-resolution images of biological structures.

Optoacoustic Effect:

The optoacoustic effect, also known as the laser-induced ultrasound or thermoacoustic effect, is a similar phenomenon to the photoacoustic effect. It occurs when a material exposed to pulsed laser light rapidly heats and expands, creating an acoustic wave. The acoustic wave carries information about the optical properties and structure of the material.

Example: In non-destructive testing, the optoacoustic effect can be used to detect flaws or structural irregularities in materials such as metals or composites. By analyzing the generated acoustic waves, engineers can identify hidden defects without damaging the material.

Acousto-optic Effect:

The acousto-optic effect describes the interaction between sound waves and light waves in a material. When an acoustic wave propagates through a material, it induces changes in the refractive index of the medium.

These changes modify the path and properties of the light waves traveling through the material.

Example: Acousto-optic devices, such as acousto-optic modulators or deflectors, utilize the acousto-optic effect to control and manipulate light. By applying an acoustic wave to a crystal, the refractive index variations induced by the wave can modulate or deflect the incident light, allowing precise control of its properties.

Synesthesia:

Synesthesia is a neurological phenomenon in which stimulation of one sensory or cognitive pathway leads to experiences in another pathway. Some individuals with synesthesia may experience the perception of colors when exposed to certain sounds or musical notes. This phenomenon highlights the intriguing cross-modal interactions between light and sound perception.

Example: A person with sound-color synesthesia may perceive specific colors or color associations when hearing particular musical notes or sounds. For example, they may consistently associate the sound of a trumpet with the color red.

Cymatics:

Cymatics is the study of the visual representation of sound waves through the use of physical mediums. It demonstrates how sound vibrations can create intricate patterns and shapes when transmitted through materials such as fluids, powders, or plates. These visual representations offer insights into the visible manifestation of sound energy.

Example: When fine sand is spread over a metal plate and subjected to specific sound frequencies, the sand forms distinct patterns known as Chladni figures. These figures demonstrate the nodal points of vibration created by the sound waves, providing a visual representation of sound energy distribution.

Luminescence:

Luminescence is the emission of light from a material that has been excited by an external energy source, such as sound. Certain materials, like phosphors or fluorescent dyes, can emit light when stimulated by acoustic waves. This phenomenon, known as sonoluminescence, occurs due to the intense compression and heating of gas bubbles within the liquid medium.

Example: In sonoluminescence experiments, researchers use powerful sound waves to create and collapse tiny gas bubbles in a liquid. The rapid compression generates high temperatures and pressures, resulting in the emission of brief flashes of light from the collapsing bubbles.

Sound-induced Light Modulation:

Sound waves can be used to modulate or control the properties of light, allowing for unique applications in optics and telecommunications. By manipulating the refractive index of a material using sound waves, the transmission, polarization, or phase of light can be altered.

Example: Acousto-optic devices, such as modulators and switches, utilize the acousto-optic effect to control the intensity or frequency of light. These devices find applications in telecommunications systems, laser beam steering, and optical signal processing.

Musical Light Shows:

In the realm of entertainment, the interaction between light and sound is often utilized to create captivating audiovisual experiences. Musical light shows combine synchronized lighting effects, such as colored lights, lasers, and visual projections, with music to enhance the emotional impact and immersiveness of a live performance or event.

Example: Concerts, music festivals, and theater productions often incorporate elaborate lighting setups that dynamically respond to the music being performed, adding a visual dimension to the auditory experience. The synchronization of lights with sound enhances the overall sensory impact and creates a more engaging and memorable event.

Sound Healing and Therapy:

The interaction between light and sound is also explored in the realm of healing and therapy. Light and sound therapies aim to stimulate relaxation, balance energy, and promote well-being. Various techniques utilize specific light and sound frequencies to induce specific physiological and psychological responses in individuals.

Example: Color light therapy involves exposing individuals to specific colors of light, often accompanied by soothing soundscapes or music. Each color is believed to have unique properties that can influence mood, emotions, and energy levels.

Cross-modal Perception:

Cross-modal perception refers to the phenomenon where stimuli from one sensory modality (such as sound) can influence or affect the perception of stimuli from another modality (such as light). The brain integrates information from different senses to create a unified perceptual experience.

Example: In multimedia art installations, artists combine visual and auditory elements to create a cohesive sensory experience. The combination of light, sound, and sometimes other sensory stimuli can evoke specific emotions, moods, or narratives, providing viewers with a multi-dimensional and immersive experience.

The interaction between light and sound encompasses a wide array of phenomena and mechanisms that showcase the interconnected nature of our senses and the fundamental principles of energy transfer. From the photoacoustic and optoacoustic effects to the ac ousto-optic effect and synesthesia, the principles and mechanisms of light-sound interaction offer profound insights into the intricate relationships between these sensory modalities.

Understanding these interactions allows us to develop innovative technologies and applications in fields such as biomedical imaging, materials testing, and sensory perception research. By harnessing the principles of light-sound interaction, we can create new avenues for exploration, expanding our knowledge of the world and deepening our understanding of the interconnected nature of our senses.

As we continue to explore the principles and mechanisms of light-sound interaction, we unlock new possibilities for cross-sensory experiences, artistic expressions, and scientific discoveries. The study of this intriguing phenomenon unveils the remarkable interplay between light and sound, enriching our understanding of the intricacies of perception and expanding the boundaries of human experience.

Photoacoustic Effect:
Generating Sound Waves with Light

The photoacoustic effect is a remarkable phenomenon that bridges the worlds of light and sound. It involves the conversion of absorbed light energy into acoustic (sound) waves, leading to a range of applications in fields such as biomedical imaging, materials characterization, and environmental monitoring. Let's delve into the principles and mechanisms of the photoacoustic effect, exploring how light can be harnessed to generate powerful sound waves and unlock valuable information.

Mechanism of the Photoacoustic Effect:

The photoacoustic effect occurs when a material absorbs light energy, leading to localized heating and rapid thermal expansion. This expansion generates pressure waves within the material, which propagate as acoustic waves that can be detected and analyzed.

The process begins with the absorption of light by a material. The absorbed photons transfer their energy to the material's molecules, increasing their kinetic energy and raising the local temperature. The temperature rise induces thermal expansion, resulting in a sudden release of pressure. This expansion and contraction cycle generates acoustic waves that can be measured and studied.

Example: In photoacoustic imaging, a pulsed laser beam is used to illuminate tissues or objects of interest. The absorbed light energy leads to localized heating, causing the material to expand and generate acoustic waves. These waves can be detected by ultrasound transducers, allowing the reconstruction of high-resolution images.

Application in Biomedical Imaging:

One of the primary applications of the photoacoustic effect is in biomedical imaging. Photoacoustic imaging combines the high contrast of optical imaging with the deep tissue penetration of ultrasound. By illuminating tissues with laser light and detecting the resulting acoustic waves, detailed images can be obtained.

This technique enables non-invasive visualization of biological structures, such as blood vessels, tumors, and organs, with excellent spatial resolution and functional information.

Example: Photoacoustic imaging has been used to detect and monitor various diseases, including cancer, cardiovascular conditions, and skin disorders. By leveraging the photoacoustic effect, clinicians can visualize tissue morphology, blood oxygenation, and biomarker distribution, aiding in early detection and treatment planning.

Material Characterization:

The photoacoustic effect also finds applications in materials science and characterization. By irradiating materials with laser pulses and analyzing the resulting acoustic waves, valuable information about the material's properties, composition, and structure can be extracted.

Different materials exhibit unique photoacoustic responses based on their absorption characteristics, enabling the identification and analysis of various substances.

Example: In pharmaceutical research, the photoacoustic effect can be used to study drug delivery systems, detect counterfeit drugs, and analyze the composition of pharmaceutical formulations. By measuring the photoacoustic signals from different materials, researchers can gain insights into their chemical composition and physical properties.

Gas Sensing and Environmental Monitoring:

The photoacoustic effect has proven valuable in gas sensing and environmental monitoring applications. Specific gases or pollutants can be detected by targeting their characteristic absorption bands with appropriate laser wavelengths.

By measuring the resulting photoacoustic signals, the presence and concentration of gases, such as carbon dioxide, methane, or volatile organic compounds, can be determined with high sensitivity and selectivity.

Example: Photoacoustic gas sensors have been employed in environmental monitoring for detecting air pollutants, assessing indoor air quality, and monitoring industrial emissions. The photoacoustic effect allows for rapid and accurate measurement of trace gas concentrations, contributing to efforts in environmental protection and safety.

Functional Imaging and Spectroscopy:

Photoacoustic imaging not only provides structural information but also offers functional insights into biological tissues. By utilizing specific wavelengths of light that are absorbed by particular molecules, researchers can map functional parameters such as blood oxygenation, blood flow, and tissue metabolism.

This functional imaging capability, coupled with the high resolution and deep tissue penetration of photoacoustic imaging, allows for comprehensive characterization of tissue physiology and can aid in the diagnosis and monitoring of diseases.

Example: Photoacoustic spectroscopy can be used to measure the oxygenation level of blood in real-time during surgical procedures, providing immediate feedback to surgeons and helping ensure adequate tissue oxygenation.

Nanoparticle Imaging and Therapy:

The photoacoustic effect can be combined with the use of light-absorbing nanoparticles to enhance imaging and therapeutic applications. Nanoparticles, such as gold nanorods or carbon nanotubes, can be selectively delivered to specific tissues or cells, enhancing the photoacoustic response and enabling targeted imaging or therapy.

For imaging, these nanoparticles act as contrast agents, amplifying the photoacoustic signal and improving the detection sensitivity. In therapy, they can be used for photothermal therapy, where laser light is converted into heat by the nanoparticles, leading to localized tissue destruction.

Example: In cancer research, photoacoustic imaging can be combined with targeted nanoparticles to visualize tumor margins and monitor the response to therapy. Additionally, photothermal therapy utilizing photoacoustic activation of nanoparticles shows promise as a localized treatment approach for cancer cells.

Photoacoustic Tomography:

Photoacoustic tomography (PAT) is an emerging imaging modality that utilizes the photoacoustic effect to reconstruct three-dimensional images of biological tissues. By acquiring photoacoustic signals from multiple angles, sophisticated algorithms can reconstruct detailed cross-sectional and volumetric images.

PAT combines the benefits of high-resolution optical imaging and deep tissue penetration of ultrasound, providing comprehensive visualization of tissue morphology and functional information.

Example: In preclinical research, PAT has been used to study the brain, breast, and other organs, enabling non-invasive imaging with high resolution and depth. PAT has the potential to advance clinical diagnostics, guiding surgical interventions, and monitoring therapeutic responses.

Photoacoustic Microscopy:

Photoacoustic microscopy (PAM) is a high-resolution imaging technique that utilizes the photoacoustic effect to visualize biological tissues at the cellular and even subcellular level. By combining optical illumination and ultrasonic detection, PAM offers label-free imaging with excellent spatial resolution.

PAM has applications in studying cellular structures, microvasculature, and functional dynamics in various biological samples.

Example: PAM has been employed to study the retina, providing detailed imaging of retinal microvasculature and assisting in the diagnosis and monitoring of retinal diseases such as diabetic retinopathy and age-related macular degeneration.

Functional Brain Imaging:

The photoacoustic effect has gained attention in neuroscience research, particularly in functional brain imaging. By utilizing light-absorbing dyes or nanoparticles that selectively target specific brain regions or cellular structures, researchers can map brain activity and connectivity with high spatial and temporal resolution.

This approach, known as functional photoacoustic imaging, offers a non-invasive and label-free method to study brain function and provides insights into neurological disorders and cognitive processes.

Example: Functional photoacoustic imaging has been used to study cerebral hemodynamics, neurovascular coupling, and brain activity patterns in animal models, offering valuable information about brain function and aiding in the development of therapeutic interventions for brain disorders.

Photoacoustic Flowmetry:

The photoacoustic effect can be utilized to measure blood flow in real-time, known as photoacoustic flowmetry. By illuminating blood vessels with laser light, the resulting photoacoustic signals can be analyzed to determine the velocity and direction of blood flow.

This technique offers advantages such as label-free imaging, high sensitivity, and the ability to visualize microcirculation, making it valuable in vascular research and clinical applications.

Example: Photoacoustic flowmetry has been used to study blood flow dynamics in various organs, including the brain, heart, and skin. It provides insights into vascular abnormalities, such as ischemia, and contributes to the understanding of cardiovascular diseases.

Photoacoustic Communication:

The photoacoustic effect can also be harnessed for communication purposes. By encoding information into modulated light signals and converting them into acoustic waves, data can be transmitted through air or water using the photoacoustic effect.

This approach offers advantages such as low interference, enhanced signal-to-noise ratio, and the ability to transmit information in environments where traditional radiofrequency communication may be challenging.

Example: Underwater communication systems have utilized the photoacoustic effect to transmit signals over long distances with minimal signal degradation. This technology finds applications in underwater exploration, marine research, and underwater data transmission.

Environmental Sensing and Pollution Monitoring:

The photoacoustic effect can be employed for environmental sensing and monitoring applications, particularly in the detection and quantification of gases and pollutants. By using specific light wavelengths that correspond to the absorption characteristics of target gases, photoacoustic spectroscopy enables accurate and sensitive gas sensing.

This approach finds applications in environmental monitoring, industrial safety, and atmospheric research.

Example: Photoacoustic spectroscopy has been used for monitoring greenhouse gases, such as carbon dioxide and methane, in the atmosphere. This enables tracking of emissions, assessing climate change impact, and supporting environmental policies.

The photoacoustic effect exemplifies the extraordinary synergy between light and sound. By harnessing the power of light to induce acoustic waves, we open up new avenues in biomedical imaging, materials characterization, and environmental sensing. The ability to convert absorbed light energy into sound waves provides valuable insights into the composition, structure, and properties of diverse materials, as well as facilitating non-invasive imaging of biological tissues. The photoacoustic effect has revolutionized the fields of medical diagnostics, materials science, and environmental monitoring, offering new possibilities for scientific research, technological advancements, and improved healthcare.

As we continue to deepen our understanding of the photoacoustic effect and refine its applications, we unlock new potentials for exploring the intricacies of the human body, characterizing materials at the microscopic level, and monitoring the environment. The harnessing of light to generate sound waves through the photoacoustic effect represents a powerful fusion of disciplines, bridging the realms of optics and acoustics to expand our knowledge, enhance our capabilities, and shape the future of numerous fields.

Acousto-Optic Effect:
Modulating Light with Sound Waves

The acousto-optic effect is a fascinating phenomenon that enables the modulation and control of light using sound waves. By exploiting the interaction between sound waves and the refractive index of a material, the acousto-optic effect offers a versatile approach to manipulate light with high precision and speed. In this overview, we delve into the principles and mechanisms of the acousto-optic effect, exploring how sound waves can be used to modulate the properties of light and unlock a multitude of applications.

Mechanism of the Acousto-Optic Effect:

The acousto-optic effect occurs when an acoustic wave propagates through a material, inducing changes in the refractive index. This change in refractive index modulates the path, phase, polarization, or intensity of the incident light. The interaction between the sound wave and the light wave takes place due to the modulation of the material's optical properties.

The acousto-optic effect can be categorized into two types: the Raman-Nath effect, which is applicable for weak acoustic waves, and the Bragg diffraction effect, which occurs when the acoustic wave intensity is high.

Example: When a sound wave propagates through a crystal or an optical medium, such as a liquid or a gas, it causes periodic variations in the refractive index of the material. This variation interacts with the incident light, leading to changes in its properties.

Acousto-Optic Modulators:

Acousto-optic modulators (AOMs) are devices that utilize the acousto-optic effect to modulate the intensity, frequency, or phase of light. AOMs consist of a crystal or a medium through which an acoustic wave is passed. The acoustic wave induces variations in the refractive index, leading to the modulation of the incident light.

AOMs offer advantages such as high modulation speeds, wide bandwidth, and precise control, making them invaluable tools in various applications, including laser technology, telecommunications, and optical signal processing.

Example: In laser systems, AOMs can be used to control the intensity of laser beams. By varying the frequency and power of the applied acoustic wave, the intensity of the diffracted or transmitted light can be precisely adjusted, allowing for laser power modulation, beam steering, or optical switching.

Acousto-Optic Deflectors:

Acousto-optic deflectors (AODs) are devices that utilize the acousto-optic effect to deflect the path of light. By applying an acoustic wave to a medium, the refractive index variations induced by the sound wave cause the incident light to change direction based on the principle of diffraction.

AODs enable fast and precise control of light beam deflection, making them valuable in applications such as laser scanning microscopy, laser beam steering, and laser printing.

Example: In laser scanning microscopy, AODs are used to rapidly deflect the laser beam across the sample, allowing for high-speed image acquisition and three-dimensional imaging. By controlling the acoustic wave frequency, the position of the laser beam can be precisely controlled, enabling detailed scanning of biological tissues or materials.

Acousto-Optic Filters:

Acousto-optic filters utilize the acousto-optic effect to control the transmission or reflection of specific wavelengths of light. By applying an acoustic wave to a material, the refractive index variations induce wavelength-dependent diffraction, leading to the selection or filtering of certain optical frequencies.

Acousto-optic filters offer advantages such as fast switching, wide tunability, and narrow bandwidth, making them valuable in spectroscopy, telecommunications, and optical signal processing.

Example: In telecommunications, acousto-optic filters are used to selectively filter or separate specific optical wavelengths in fiber optic communication systems. By adjusting the frequency of the acoustic wave, different wavelengths can be transmitted or blocked, enabling efficient wavelength division multiplexing (WDM) and optical signal routing.

Acousto-Optic Modulation in Spectroscopy:

The acousto-optic effect finds extensive applications in spectroscopy, enabling precise control and manipulation of light for analytical purposes. By modulating the intensity, frequency, or phase of light using an acousto-optic modulator, spectroscopic measurements can be performed with high accuracy and sensitivity.

Acousto-optic modulation in spectroscopy techniques such as Raman spectroscopy, Fourier-transform infrared spectroscopy (FTIR), and fluorescence spectroscopy enhances the signal-to-noise ratio, improves spectral resolution, and enables rapid data acquisition.

Example: In Raman spectroscopy, an acousto-optic modulator is used to select and modulate specific excitation wavelengths. This allows for precise targeting of molecular vibrational modes, enhancing the sensitivity and selectivity of Raman measurements.

Acousto-Optic Tunable Filters:

Acousto-optic tunable filters (AOTFs) exploit the acousto-optic effect to provide wavelength tunability and spectral control. By adjusting the frequency of the applied acoustic wave, the transmission or reflection properties of the AOTF can be dynamically tuned, enabling rapid and precise spectral filtering.

AOTFs find applications in hyperspectral imaging, fluorescence microscopy, spectroscopy, and remote sensing, providing versatile tools for spectral analysis and selective imaging.

Example: In hyperspectral imaging, AOTFs are employed to rapidly switch between different spectral bands, allowing the acquisition of multispectral or hyperspectral images. This enables detailed analysis and identification of materials based on their spectral signatures.

Acousto-Optic Data Processing:

The acousto-optic effect offers opportunities for data processing and manipulation in optical communication and signal processing systems. By modulating the properties of light using sound waves, functionalities such as signal modulation, frequency shifting, and wavelength conversion can be achieved.

This capability enables advanced optical signal processing, including frequency-domain filtering, optical time-division multiplexing (OTDM), and coherent communication systems.

Example: In coherent optical communication systems, acousto-optic modulators are utilized for frequency shifting and phase modulation of optical signals. This allows for signal processing techniques such as quadrature amplitude modulation (QAM), dispersion compensation, and coherent detection, enhancing the capacity and performance of high-speed optical communication networks.

Acousto-Optic Holography:

The acousto-optic effect plays a significant role in acousto-optic holography, a technique that combines sound waves and light waves to create holographic images. By utilizing an acousto-optic modulator, the phase and amplitude of the reference and object beams in holography can be precisely controlled, enabling the reconstruction of three-dimensional images.

Acousto-optic holography finds applications in holographic displays, optical storage, and digital holography, providing a powerful means to visualize complex three-dimensional structures.

Example: Acousto-optic holography has been used to create holographic displays, where dynamic three-dimensional images can be generated by controlling the acoustic wave parameters. This technology has the potential to revolutionize visual communication and immersive displays.

Acousto-Optic Beam Steering:

The acousto-optic effect allows for precise control of the direction and angle of a light beam through acousto-optic deflectors. By adjusting the frequency and power of the acoustic wave, the deflection angle of the incident light can be controlled, enabling rapid and accurate beam steering.

Acousto-optic beam steering finds applications in laser scanning systems, laser radar, free-space optical communication, and optical instrumentation.

Example: In laser radar systems, acousto-optic deflectors are utilized to steer laser beams for target tracking and range-finding applications. The ability to rapidly change the direction of the laser beam allows for efficient scanning of the surrounding environment.

Acousto-Optic Frequency Shifters:

The acousto-optic effect can be employed for frequency shifting of light signals. By applying an acoustic wave to a medium, the frequency of the incident light can be altered based on the acousto-optic interaction, enabling frequency conversion or shifting of optical signals.

Acousto-optic frequency shifters find applications in optical communications, coherent detection systems, and spectroscopy.

Example: In optical communication systems, acousto-optic frequency shifters are used for frequency modulation or demodulation of optical signals. This facilitates multiplexing and demultiplexing of optical channels and enables compatibility with different communication standards.

Acousto-Optic Spatial Light Modulators:

Acousto-optic spatial light modulators (AOSLMs) utilize the acousto-optic effect to dynamically control the spatial intensity or phase profile of light. By applying an acoustic wave to a medium, the transmitted or reflected light can be spatially modulated, allowing for precise manipulation of optical wavefronts.

AOSLMs find applications in wavefront shaping, adaptive optics, holography, and optical information processing.

Example: In adaptive optics systems, AOSLMs are used to compensate for wavefront distortions caused by atmospheric turbulence or optical aberrations. By dynamically adjusting the phase of the incident light, AOSLMs enable real-time correction, improving the resolution and quality of imaging and laser systems.

Acousto-Optic Time-Domain Reflectometers:

Acousto-optic time-domain reflectometers (AOTDRs) utilize the acousto-optic effect to perform high-resolution measurements of the length and characteristics of optical fibers. By generating an acoustic wave along the fiber and analyzing the reflected optical signal, AOTDRs can accurately detect and locate fiber faults, discontinuities, or losses.

AOTDRs find applications in fiber optic network troubleshooting, maintenance, and quality control.

Example: In telecommunications, AOTDRs are used to identify and locate faults or degradation in optical fiber links. By analyzing the backscattered light signals, AOTDRs provide valuable information about the condition and quality of the fiber, assisting in rapid fault diagnosis and repair.

The acousto-optic effect provides a powerful means to modulate and control light using sound waves. Through the interaction between sound waves and the refractive index of a material, the acousto-optic effect enables precise manipulation of light properties such as intensity, frequency, phase, and direction. Acousto-optic devices, including modulators, deflectors, and filters, find applications in laser technology, telecommunications, spectroscopy, and optical signal processing.

By harnessing the acousto-optic effect, we unlock new possibilities for high-speed data communication, advanced imaging techniques, and optical control systems. The integration of sound waves and light waves through the acousto-optic effect showcases the synergistic relationship between different forms of energy and expands our ability to shape and harness light for diverse applications.

V.   Optical Techniques for Sound Detection

Traditional methods of sound detection often rely on the use of microphones or other mechanical sensors. However, optical techniques offer a unique and powerful alternative, allowing for highly sensitive and non-contact detection of sound waves. By harnessing light and its interaction with materials, optical techniques enable precise and versatile sound detection across a wide range of applications. Here are some various optical techniques for sound detection, providing examples and explanations of their principles and applications.

Laser Doppler Vibrometry:

Laser Doppler vibrometry (LDV) is an optical technique used for precise measurement of vibrations induced by sound waves. LDV relies on the principle of Doppler shift, where the frequency of light scattered from a vibrating surface is shifted in proportion to the velocity of the surface motion.

LDV systems utilize laser light to illuminate the vibrating object, and the scattered light is detected and analyzed to measure the velocity and displacement of the surface. This enables non-contact and high-resolution sound detection.

Example: In structural engineering, LDV can be used to measure the vibrations of buildings, bridges, or other structures induced by sound or environmental factors. It provides valuable information about structural integrity and behavior, aiding in structural monitoring and maintenance.

Fiber Optic Sensors:

Fiber optic sensors offer a versatile and sensitive approach for sound detection. By utilizing the optical properties of optical fibers, changes in acoustic pressure or vibrations can be converted into changes in light intensity, phase, or wavelength.

Various fiber optic sensor configurations, such as Fabry-Perot interferometers, Bragg gratings, or microbend sensors, can be employed to detect sound waves. These sensors can be integrated into structures or deployed in harsh environments for remote or distributed sound sensing.

Example: In underwater acoustics, fiber optic hydrophones are used for sound detection in marine environments. By using fiber optic cables with hydrophone transducers, sound waves can be converted into changes in light intensity, allowing for precise underwater sound measurements in applications such as underwater communication or environmental monitoring.

Laser-Interferometric Techniques:

Laser-interferometric techniques leverage the interference of laser light to detect and measure sound waves. These techniques exploit the modulation of the optical path length induced by sound-induced vibrations, allowing for highly sensitive sound detection.

Common laser-interferometric techniques for sound detection include laser interferometers, Michelson interferometers, and Fabry-Perot interferometers. These systems utilize the interference patterns of laser light to detect and analyze the phase or intensity changes caused by sound waves.

Example: In ultrasonic imaging, laser interferometry can be used to visualize and characterize acoustic waves propagating through materials. By measuring the phase shifts or intensity variations caused by the sound waves, detailed images and information about material properties can be obtained.

Photorefractive Techniques:

Photorefractive materials provide a unique approach for sound detection through their ability to record and amplify sound-induced optical changes. These materials exhibit a photorefractive effect, where changes in refractive index are induced by light intensity variations, including those caused by sound waves.

Photorefractive techniques utilize the interaction between sound waves and the refractive index changes in these materials, allowing for the non-contact detection and visualization of sound waves.

Example: In non-destructive testing, photorefractive techniques can be used to visualize acoustic waves propagating through materials. By illuminating the material with laser light and detecting the changes in refractive index induced by sound waves, high-resolution images of internal structures and defects can be obtained.

Optical Microphone Arrays:

Optical microphone arrays consist of multiple optical sensors strategically arranged to capture sound waves from different directions. Each sensor detects the changes in light intensity or phase caused by sound waves, allowing for precise sound source localization and beamforming.

Optical microphone arrays offer advantages such as high spatial resolution, low cross-talk, and the ability to capture sound in challenging environments with high ambient noise.

Example: Optical microphone arrays find applications in various fields, including teleconferencing, voice recognition, and surveillance systems. They enable accurate sound source localization, noise cancellation, and enhanced speech intelligibility in complex acoustic environments.

Brillouin Scattering:

Brillouin scattering is an optical phenomenon that enables the detection and characterization of sound waves in materials. It involves the interaction between light and acoustic waves, resulting in a shift in the frequency of scattered light proportional to the sound wave frequency.

Brillouin scattering techniques, such as Brillouin optical time-domain reflectometry (BOTDR) or Brillouin optical time-domain analysis (BOTDA), offer distributed sensing capabilities for measuring sound-induced strain or temperature changes along an optical fiber.

Example: Brillouin scattering techniques have applications in structural health monitoring, geophysics, and energy exploration. They allow for the detection of sound waves and the monitoring of structural vibrations or changes in the environment over long distances.

Heterodyne Detection:

Heterodyne detection is a technique that utilizes the interference between two light waves of slightly different frequencies to detect sound waves. The interaction between the sound wave and the optical carrier frequencies creates a beat frequency that can be detected and analyzed.

Heterodyne detection offers high sensitivity and precise measurement capabilities, making it suitable for applications requiring ultra-low noise detection.

Example: Heterodyne detection is used in applications such as laser spectroscopy, optical communications, and ultrasound imaging. It allows for highly sensitive detection of sound waves or acoustic signals in various frequency ranges.

Photoelasticity:

Photoelasticity is an optical technique that utilizes the birefringent properties of materials to visualize and analyze sound-induced stress or strain patterns. By illuminating a photoelastic material with polarized light, changes in the polarization state indicate the distribution of stress or strain caused by sound waves.

Photoelasticity offers insights into the mechanical behavior of materials under acoustic loading, enabling the analysis of stress distributions and the detection of acoustic emission.

Example: Photoelasticity is used in the study of structural mechanics, material characterization, and the design of acoustic devices. It helps visualize and analyze stress patterns in objects subjected to sound waves or mechanical vibrations.

Coherent Phonon Spectroscopy:

Coherent phonon spectroscopy is an optical technique that allows for the detection and characterization of acoustic vibrations in materials. By using ultrafast laser pulses, coherent phonons—collective vibrations of atoms or molecules—can be generated and probed to obtain information about the material's acoustic properties.

Coherent phonon spectroscopy offers insights into phonon lifetimes, phonon-phonon interactions, and phonon dispersion, providing valuable information about the material's mechanical behavior and thermal properties.

Example: Coherent phonon spectroscopy is utilized in the study of materials with applications in energy storage, thermoelectrics, and phononic devices. It enables the characterization of acoustic properties and aids in the development of materials with tailored acoustic behavior.

Optoacoustic Imaging:

Optoacoustic imaging, also known as photoacoustic imaging, combines optical excitation and acoustic detection to create detailed images of biological tissues. By illuminating the tissue with short laser pulses, sound waves are generated due to the photoacoustic effect, and these waves are detected to reconstruct images.

Optoacoustic imaging offers deep tissue penetration, high spatial resolution, and functional imaging capabilities, making it valuable in biomedical research, cancer detection, and monitoring of physiological parameters.

Example: Optoacoustic imaging has been used in preclinical and clinical settings to visualize tumor vasculature, study brain activity, and monitor oxygenation levels in tissues. It provides valuable information for disease diagnosis, treatment planning, and understanding biological processes.

Interferometric Fiber Optic Sensors:

Interferometric fiber optic sensors utilize the interference of light waves within optical fibers to detect sound-induced vibrations or strain changes. By exploiting the phase shifts or intensity variations caused by acoustic waves, these sensors provide high sensitivity and precise measurements.

Interferometric fiber optic sensors can be configured as Fabry-Perot interferometers, Mach-Zehnder interferometers, or Michelson interferometers, offering flexibility and versatility in different sensing applications.

Example: Interferometric fiber optic sensors find applications in structural health monitoring, seismic sensing, and vibration analysis. They allow for the detection of acoustic waves or vibrations with high sensitivity and can be integrated into structures for continuous monitoring.

Acoustic Optical Coherence Tomography:

Acoustic optical coherence tomography (AOCT) combines the principles of optical coherence tomography (OCT) with acoustic sensing to image tissues and detect sound-induced vibrations. By utilizing low-coherence interferometry, AOCT can provide depth-resolved imaging while simultaneously detecting acoustic waves.

AOCT offers the advantages of high-resolution imaging and the ability to visualize tissue morphology and assess biomechanical properties.

Example: AOCT has applications in ophthalmology for imaging the retina and assessing its mechanical properties. It enables the diagnosis and monitoring of retinal diseases and aids in the understanding of ocular biomechanics.

Raman Spectroscopy:

Raman spectroscopy, a widely used optical technique, can also be employed for sound detection. When sound waves interact with a material, they can induce changes in the Raman scattering of light, providing information about the material's vibrational properties and sound-induced stress.

Raman spectroscopy-based sound detection offers insights into the mechanical behavior of materials, structural dynamics, and acoustic phenomena.

Example: Raman spectroscopy has been utilized to study sound-induced vibrations in materials such as graphene, nanowires, and thin films. It enables the characterization of acoustic properties at the micro- and nanoscale, contributing to the understanding of material behavior and the development of advanced materials.

Optical Microresonators:

Optical microresonators are small devices that confine light within a resonant cavity, allowing for highly sensitive detection of sound waves. When sound waves interact with the microresonator, they induce changes in the resonance frequency or quality factor, which can be measured optically.

Optical microresonators offer advantages such as high sensitivity, compact size, and compatibility with integrated photonic circuits.

Example: Optical microresonators have applications in ultrasound sensing, environmental monitoring, and biosensing. They enable the detection of sound waves with high sensitivity and can be integrated into wearable or implantable devices for various sensing applications.

Coherent Optical Time-Domain Reflectometry:

Coherent optical time-domain reflectometry (C-OTDR) is an optical technique that utilizes backscattered light to detect and locate sound-induced vibrations along an optical fiber. By analyzing the phase or intensity changes of the backscattered light, C-OTDR provides information about the acoustic disturbances along the fiber.

C-OTDR offers distributed sensing capabilities, allowing for the monitoring of long optical fibers with high spatial resolution.

Example: C-OTDR has applications in structural health monitoring of civil infrastructure, pipeline monitoring, and perimeter security. It enables the detection of acoustic events, such as intrusion or impact, along the length of the optical fiber.

Optical Coherence Elastography:

Optical coherence elastography (OCE) combines optical coherence tomography (OCT) with mechanical excitation to detect and image sound-induced mechanical waves in tissues. By applying acoustic waves to the tissue and measuring the resulting mechanical displacements with OCT, OCE provides information about tissue elasticity and mechanical properties.

OCE finds applications in biomedical research, tissue characterization, and early detection of diseases.

Example: OCE has been used to study the biomechanical properties of tissues, such as the cornea, skin, and cardiovascular tissues. It enables the assessment of tissue stiffness and can aid in the diagnosis and monitoring of diseases like cancer or cardiovascular disorders.

Optical techniques for sound detection offer unique advantages in terms of sensitivity, non-contact operation, and versatility. By harnessing the interaction between light and sound waves, these techniques facilitate precise and advanced sound detection across a wide range of applications.

By leveraging the unique properties of light and its interaction with materials, optical techniques for sound detection open up new possibilities for scientific research, engineering applications, and industrial processes. They offer valuable insights into the behavior of sound waves, the dynamics of structures, and the characteristics of materials.

Continued advancements in optical technologies, including the development of novel sensing configurations, improved signal processing algorithms, and miniaturization of devices, will further enhance the capabilities and applications of optical techniques for sound detection. These techniques play a vital role in expanding our understanding of sound-related phenomena and driving innovations in various fields where precise sound detection is crucial.

The integration of light-based technologies and sound detection enables us to explore sound waves in new ways, providing valuable information and opening doors to innovative solutions. Optical techniques for sound detection empower us to push the boundaries of scientific exploration, engineering applications, and technological advancements, ultimately contributing to a better understanding of our acoustic environment and enhancing our ability to harness sound for a variety of purposes.

Laser Interferometry:
Measuring Sound Waves with Light

Laser interferometry is a powerful optical technique that allows for the precise measurement of sound waves by leveraging the interference of light. By utilizing the principles of interference and phase modulation, laser interferometry enables accurate detection, characterization, and analysis of sound-induced vibrations. Now to explore the principles, configurations, and applications of laser interferometry for measuring sound waves with exceptional precision.

Interference and Phase Modulation:

At the core of laser interferometry lies the principle of interference—the interaction of light waves that results in constructive or destructive interference. By splitting a laser beam into two paths—one acting as the reference and the other interacting with the sound-induced vibrations—laser interferometry detects changes in the interference pattern, which provide information about the sound waves.

Phase modulation is employed to introduce a phase shift in the reference or signal beam based on the vibrations being measured. This phase shift is detected and analyzed to extract information about the sound wave characteristics, including frequency, amplitude, and phase.

Example: In a Michelson interferometer configuration, a laser beam is split into two paths—one is reflected by a fixed mirror (reference beam), and the other is directed towards a vibrating surface (signal beam). When the two beams are recombined, their interference pattern is modified due to the phase difference caused by the sound-induced vibrations. This pattern is captured and analyzed to determine the properties of the sound waves.

Doppler Laser Vibrometry:

Doppler laser vibrometry is a specific application of laser interferometry that utilizes the Doppler effect to measure sound-induced vibrations. When the sound wave interacts with a vibrating surface, it induces changes in the frequency of the reflected laser light, which are proportional to the surface velocity.

By analyzing the frequency shift of the reflected light, Doppler laser vibrometry provides valuable information about the sound wave's frequency, velocity, and displacement.

Example: In laser Doppler vibrometry, a laser beam is directed towards a vibrating surface. The reflected beam undergoes a frequency shift due to the Doppler effect caused by the sound-induced vibrations. By analyzing this frequency shift, the amplitude and frequency of the sound waves can be determined. This technique finds applications in areas such as structural analysis, quality control, and musical instrument testing.

Fiber-Optic Interferometry:

Fiber-optic interferometry utilizes optical fibers to guide the laser light, enabling remote and distributed sound wave measurements. By incorporating optical fibers with interferometric sensors or Fabry-Perot cavities, sound-induced vibrations can be detected and measured at different points along the fiber length.

Fiber-optic interferometry offers advantages such as high sensitivity, immunity to electromagnetic interference, and the ability to monitor sound waves over long distances.

Example: In structural health monitoring, fiber-optic interferometry can be used to measure sound-induced vibrations or detect structural anomalies. By embedding fiber-optic sensors in critical areas of structures such as bridges or pipelines, the vibrations caused by sound waves or mechanical stress can be monitored remotely and continuously.

Digital Holography:

Digital holography combines laser interferometry with digital imaging techniques to capture and reconstruct sound-induced vibrations in real-time. By recording the interference pattern between a reference beam and a beam scattered by the sound wave, a digital hologram is obtained, providing a full-field representation of the sound wave's amplitude and phase.

Digital holography enables high-resolution visualization and analysis of sound waves, offering insights into their spatial distribution and behavior.

Example: Digital holography has been used to study sound propagation in various applications, including aeroacoustics, underwater acoustics, and ultrasonics. By capturing and reconstructing the holograms, researchers can analyze the spatial characteristics of sound waves, visualize acoustic phenomena, and investigate their interactions with different materials or environments.

Scanning Laser Doppler Vibrometry:

Scanning laser Doppler vibrometry (SLDV) combines laser interferometry with scanning technology to map sound-induced vibrations over a surface. By scanning the laser beam across the surface of an object, SLDV captures the vibrations at multiple points, allowing for the visualization and analysis of the spatial distribution of sound waves.

SLDV offers high-resolution imaging capabilities, making it suitable for applications such as modal analysis, defect detection, and non-destructive testing.

Example: In the automotive industry, SLDV is used to analyze the vibrations of vehicle components caused by sound waves or mechanical forces. By scanning the laser beam across critical parts, such as engine components or body panels, the spatial characteristics and resonant frequencies can be determined, aiding in design optimization and quality control.

Heterodyne Interferometry:

Heterodyne interferometry is a technique that combines two laser beams of slightly different frequencies to measure sound-induced vibrations. By introducing a frequency difference between the reference and signal beams, the interference pattern generated contains information about the sound waves' phase and amplitude.

Heterodyne interferometry offers high sensitivity and accuracy, allowing for precise measurement of sound waves in various environments and materials.

Example: In material characterization, heterodyne interferometry can be employed to measure the mechanical properties and viscoelastic behavior of materials under the influence of sound waves. The technique enables the determination of material parameters such as Young's modulus, damping coefficients, and acoustic impedance.

Laser Ultrasonics:

Laser ultrasonics combines laser interferometry with the generation and detection of ultrasonic waves to measure sound-induced vibrations. A laser pulse is used to generate ultrasonic waves in a sample, and another laser beam is employed to detect the resulting vibrations.

Laser ultrasonics offers advantages such as non-contact operation, high spatial resolution, and the ability to measure sound waves in challenging environments or on delicate surfaces.

Example: In non-destructive testing, laser ultrasonics is used to inspect materials for defects or assess their integrity. By analyzing the characteristics of the ultrasonic waves generated by sound waves, hidden cracks, delaminations, or structural abnormalities can be detected.

Quantum Cascade Laser Interferometry:

Quantum cascade laser (QCL) interferometry is an emerging technique that utilizes QCLs as light sources to measure sound-induced vibrations. QCLs emit mid-infrared light, allowing for precise and sensitive detection of sound waves in the frequency range where many gases exhibit unique absorption features.

QCL interferometry offers the potential for gas sensing, environmental monitoring, and industrial applications where specific gases or chemical species need to be detected.

Example: In environmental monitoring, QCL interferometry can be used to measure sound waves generated by gas leaks or emissions. By analyzing the interference patterns generated by the interaction of the QCL light with the gas-induced vibrations, the presence and concentration of specific gases can be determined.

Laser Doppler Acoustic Microscopy:

Laser Doppler acoustic microscopy (LDAM) combines laser interferometry with acoustic microscopy to visualize and measure sound waves with high spatial resolution. By focusing a laser beam onto a sample, the acoustic waves generated by the sample's vibrations are detected using interferometric techniques.

LDAM enables the mapping of sound wave propagation and provides information about the local mechanical properties of materials.

Example: LDAM is used in materials science and biology to study the acoustic properties of small-scale structures, such as microelectromechanical systems (MEMS) devices or biological cells. It allows for the visualization and analysis of acoustic phenomena at the microscale.

Time-Reversal Acoustics:

Time-reversal acoustics is a technique that utilizes laser interferometry to focus and manipulate sound waves. By recording the interference pattern of sound waves with a reference beam, the phase and amplitude information is stored. This recorded information can be re-emitted by the interferometer, effectively focusing the sound waves to their original source.

Time-reversal acoustics enables the selective focusing and direction of sound waves, providing control over acoustic fields.

Example: Time-reversal acoustics has applications in various fields, including medical imaging, underwater acoustics, and communications. It allows for the focusing of sound waves to specific regions of interest, improving imaging resolution, communication range, and underwater localization.

Laser-Generated Ultrasound:

Laser-generated ultrasound (LGU) is a technique that uses laser-induced thermoelastic expansion to generate ultrasonic waves for sound wave measurement. A laser beam is absorbed by a material, leading to localized heating and subsequent expansion, which generates ultrasonic waves.

By detecting and analyzing the generated ultrasound waves with laser interferometry, information about the sound waves, including their frequency and amplitude, can be obtained.

Example: LGU finds applications in non-destructive testing, material characterization, and biomedical imaging. It enables the detection of internal flaws, such as cracks or voids, in materials or the imaging of biological tissues with high resolution.

Dual-Wavelength Interferometry:

Dual-wavelength interferometry utilizes two laser beams of different wavelengths to measure sound-induced vibrations. The interference pattern formed by the two beams provides information about the sound waves, including their displacement and frequency.

Dual-wavelength interferometry offers advantages such as high sensitivity, immunity to environmental noise, and the ability to measure large amplitudes.

Example: Dual-wavelength interferometry is used in various applications, including vibration analysis, seismic monitoring, and precision engineering. It enables the measurement of sound waves with high accuracy and allows for the detection of tiny displacements or vibrations in mechanical systems.

Laser interferometry stands as a remarkable optical technique for measuring sound waves with precision and accuracy. By leveraging the principles of interference and phase modulation, laser interferometry allows us to capture, analyze, and understand the properties of sound-induced vibrations.

From Doppler laser vibrometry to fiber-optic interferometry and digital holography, laser interferometry finds applications in a wide range of fields, including structural analysis, quality control, remote sensing, and scientific research. These techniques provide invaluable insights into the behavior of sound waves, aiding in the design and optimization of acoustic systems, the evaluation of structural integrity, and the advancement of various scientific disciplines.

As technology continues to advance, laser interferometry techniques are being refined and integrated into more complex systems, allowing for multi-dimensional measurements, higher spatial resolutions, and real-time monitoring. This paves the way for further innovations and discoveries in the realm of sound wave measurement and analysis.

Laser interferometry, with its ability to harness light to measure sound waves, exemplifies the synergy between optics and acoustics. It enables us to explore the intricacies of sound phenomena, deepen our understanding of acoustic behavior, and push the boundaries of scientific exploration and technological advancements.

Fiber Optic Sensors:
Transducing Sound to Optical Signals

Fiber optic sensors have emerged as powerful transducers that convert sound waves into optical signals, revolutionizing the field of acoustic sensing. By exploiting the unique properties of optical fibers, such as their low loss, high sensitivity, and immunity to electromagnetic interference, fiber optic sensors enable highly accurate and reliable measurement of sound-induced vibrations. Now, into the principles, configurations, and applications of fiber optic sensors for transducing sound to optical signals.

Fabry-Perot Interferometric Sensors:

Fabry-Perot interferometric sensors employ a small air cavity or a thin diaphragm at the end of an optical fiber. Sound waves cause the cavity or diaphragm to vibrate, modulating the optical signal propagating through the fiber.

By analyzing changes in the optical interference pattern, the amplitude, frequency, and phase of the sound waves can be determined with exceptional accuracy.

Example: In structural health monitoring, Fabry-Perot interferometric sensors can be embedded in buildings, bridges, or aircraft to detect and monitor vibrations induced by sound waves or mechanical stress. These sensors provide valuable insights into structural integrity and behavior.

Fiber Bragg Grating Sensors:

Fiber Bragg grating (FBG) sensors consist of a periodic refractive index modulation within an optical fiber. When sound waves deform the fiber, the grating experiences strain, leading to a shift in the reflected wavelength of light.

By monitoring the wavelength shift, the characteristics of sound waves, such as their amplitude and frequency, can be precisely measured.

Example: In underwater acoustics, FBG sensors are utilized to detect and monitor sound waves in marine environments. These sensors offer advantages such as high sensitivity, long-term stability, and resistance to harsh underwater conditions.

Polarimetric Fiber Optic Sensors:

Polarimetric fiber optic sensors exploit the changes in the polarization state of light induced by sound waves. By analyzing the changes in polarization caused by the interaction of the sound waves with a sensing element, information about the sound wave's characteristics can be obtained.

Polarimetric sensors offer high sensitivity and can be used for distributed sensing along the length of the fiber.

Example: In geophysics, polarimetric fiber optic sensors can be deployed in the ground to detect and monitor seismic waves generated by natural events or human activities. These sensors provide valuable data for earthquake detection, structural monitoring, and oil and gas exploration.

Sagnac Interferometric Sensors:

Sagnac interferometric sensors exploit the Sagnac effect, which is the phase shift experienced by light propagating in a looped optical fiber. Sound waves perturb the fiber, causing a phase change in the light circulating within the loop.

By analyzing the phase shift, the sound wave's characteristics, including amplitude and frequency, can be determined.

Example: In aerospace applications, Sagnac interferometric sensors can be used to measure sound-induced vibrations in aircraft structures. These sensors offer high sensitivity, immunity to electromagnetic interference, and the ability to withstand harsh environmental conditions.

Intensity-Based Fiber Optic Sensors:

Intensity-based fiber optic sensors measure changes in light intensity induced by sound waves. These sensors utilize various mechanisms such as microbending, macrobending, or acoustic-induced changes in fiber properties to modulate the light intensity.

By monitoring the intensity variations, information about the sound wave's amplitude and frequency can be extracted.

Example: In industrial noise monitoring, intensity-based fiber optic sensors can be deployed in factories or manufacturing plants to assess noise levels and detect sound-induced vibrations. These sensors provide real-time data for noise control and worker safety.

Distributed Acoustic Sensing (DAS):

Distributed acoustic sensing (DAS) is a technique that utilizes fiber optic cables as distributed sensors for monitoring sound waves along their entire length. By analyzing the backscattered or Rayleigh scattered light in the fiber, DAS can detect and locate sound-induced vibrations at different points along the cable.

DAS offers advantages such as high spatial resolution, long sensing range, and the ability to monitor multiple events simultaneously.

Example: In oil and gas exploration, DAS is used to monitor seismic activity, detect fluid flow, and identify wellbore integrity issues. By deploying fiber optic cables along the wellbore or within the reservoir, DAS provides real-time monitoring of sound-induced vibrations for improved reservoir characterization and production optimization.

Interferometric Optical Time-Domain Reflectometry (OTDR):

Interferometric optical time-domain reflectometry (OTDR) is a technique that utilizes fiber optic cables to detect sound-induced vibrations based on changes in backscattered light. By analyzing the reflected optical signals along the fiber, OTDR can measure sound wave characteristics, such as amplitude and frequency, at different locations.

OTDR offers advantages such as long-range sensing, high sensitivity, and the ability to monitor dynamic events.

Example: In structural monitoring, OTDR can be used to detect and locate sound-induced vibrations in bridges, dams, or pipelines. By installing fiber optic cables along the structures, OTDR provides continuous monitoring and early detection of structural damage or abnormal vibrations.

Acoustic Emission Sensing:

Acoustic emission (AE) sensing utilizes fiber optic sensors to detect and measure sound waves generated by the rapid release of elastic energy in materials. By capturing the acoustic signals using fiber optic sensors,

AE sensing enables the identification of defect formation, crack propagation, or material failure.

AE sensing offers advantages such as high sensitivity, real-time monitoring, and the ability to detect and locate events with precision.

Example: In structural integrity assessment, AE sensing can be used to monitor critical components in infrastructure, such as bridges or pressure vessels. By placing fiber optic sensors at strategic locations, AE sensing provides early warning of structural damage or fatigue, enabling timely maintenance or repair.

Biomedical Applications:

Fiber optic sensors have found applications in biomedical settings for transducing sound to optical signals. For example, in intravascular ultrasound (IVUS), fiber optic sensors are used to detect and measure sound waves for imaging blood vessels and assessing plaque buildup.

Additionally, fiber optic sensors have been employed in hearing aids to convert sound waves into optical signals, allowing for enhanced signal processing and improved sound amplification.

Example: In biomedical imaging, fiber optic sensors can be used in endoscopy procedures for imaging internal organs or tissues. By incorporating fiber optic sensors into the endoscope, the acoustic signals can be transduced into optical signals for precise imaging and diagnosis.

Hydrophone Arrays:

Hydrophone arrays utilize fiber optic sensors for underwater sound detection and localization. By integrating fiber optic sensors into an array configuration, it is possible to accurately measure the direction, amplitude, and frequency of underwater sound waves.

Hydrophone arrays offer advantages such as high sensitivity, wide frequency response, and the ability to distinguish different sources of sound.

Example: In underwater acoustics research, hydrophone arrays with fiber optic sensors are used to study marine mammal communication, underwater noise pollution, and sonar systems. These arrays provide valuable data for understanding marine ecosystems, enhancing underwater communication, and monitoring underwater noise levels.

Vibration Monitoring in Machinery:

Fiber optic sensors can be employed for vibration monitoring in machinery and rotating equipment. By attaching fiber optic sensors to critical components, such as bearings or shafts, sound-induced vibrations can be transduced into optical signals for analysis and condition monitoring.

Vibration monitoring with fiber optic sensors offers benefits such as high accuracy, real-time data, and the ability to operate in harsh environments.

Example: In industrial applications, fiber optic sensors are used for vibration monitoring in turbines, motors, and rotating machinery. By detecting and analyzing sound-induced vibrations, potential faults or imbalances can be identified early, allowing for preventive maintenance and reducing downtime.

Sonar Systems:

Sonar systems utilize fiber optic sensors for underwater sound detection and imaging. By converting sound waves into optical signals, fiber optic sensors enable high-resolution imaging and accurate measurement of underwater objects, distances, and depths.

Sonar systems with fiber optic sensors offer advantages such as wide bandwidth, low noise, and the ability to operate in challenging underwater environments.

Example: Sonar systems with fiber optic sensors are utilized in various applications, including marine navigation, fishery research, and underwater mapping. These systems provide detailed information about underwater topography, object detection, and marine ecosystem monitoring.

Structural Acoustic Monitoring:

Fiber optic sensors can be integrated into structural materials to enable continuous monitoring of sound-induced vibrations and acoustic behavior. By embedding fiber optic sensors in materials such as concrete, composites, or metals, changes in the acoustic response can be detected and analyzed.

Structural acoustic monitoring with fiber optic sensors offers benefits such as distributed sensing, long-term stability, and the ability to monitor large areas.

Example: In civil engineering, fiber optic sensors are embedded in concrete structures, bridges, or tunnels to monitor sound-induced vibrations caused by environmental factors or traffic. These sensors provide valuable data for structural health monitoring, ensuring the safety and integrity of critical infrastructure.

Fiber optic sensors have transformed acoustic sensing by transducing sound waves into optical signals with unparalleled precision. Through configurations such as Fabry-Perot interferometric sensors, Fiber Bragg grating sensors, polarimetric sensors, Sagnac interferometric sensors, and intensity-based sensors, fiber optic sensors offer a wide range of options for accurately measuring sound-induced vibrations.

The applications of fiber optic sensors span across various industries and fields. They are used in structural health monitoring to detect vibrations in buildings and bridges, in underwater acoustics for monitoring sound waves in marine environments, in geophysics for seismic monitoring, in aerospace for measuring vibrations in aircraft structures, and in industrial settings for noise monitoring and control.

The key advantages of fiber optic sensors include their high sensitivity, immunity to electromagnetic interference, compact size, and suitability for distributed sensing over long distances. They provide reliable and real-time data, allowing for precise monitoring, analysis, and control of sound-induced vibrations.

As fiber optic technology continues to advance, fiber optic sensors are becoming even more sophisticated. Ongoing research and development focus on improving their sensitivity, multiplexing capabilities, and compatibility with harsh environments.

Fiber optic sensors offer a powerful and versatile solution for transducing sound into optical signals. With their exceptional precision and reliability, they contribute to enhanced understanding, monitoring, and control of acoustic phenomena in diverse applications. As technology progresses, fiber optic sensors will continue to play a vital role in driving advancements in acoustic sensing, enabling us to explore and utilize sound waves in innovative ways.

VI.   Applications of Light-Sound Transduction

The transduction of light to sound and vice versa has opened up a world of possibilities in numerous fields, enabling innovative applications that leverage the interactions between these two fundamental forms of energy. In this overview, we explore some of the remarkable applications of light-sound transduction, showcasing the diverse range of disciplines that benefit from this synergy.

Acoustic Imaging and Holography:

Light-sound transduction plays a pivotal role in acoustic imaging and holography, enabling the visualization and reconstruction of sound waves in a spatially resolved manner. By converting sound waves into optical signals, detailed images and holograms of acoustic phenomena can be captured and analyzed.

Example: In medical imaging, photoacoustic imaging combines laser-induced sound waves with optical detection to create detailed images of biological tissues. It allows for non-invasive visualization of deep tissue structures, aiding in the diagnosis and monitoring of diseases such as cancer.

Non-Destructive Testing (NDT):

Light-sound transduction finds extensive use in non-destructive testing applications, where it facilitates the inspection and characterization of materials without causing damage. By employing various optical techniques, sound waves can be generated, detected, and analyzed to assess the integrity and quality of objects.

Example: In ultrasonic testing, laser-generated ultrasound and laser interferometry are employed to detect defects, measure material properties, and evaluate the structural soundness of components in industries such as aerospace, automotive, and manufacturing.

Sensing and Monitoring:

Light-sound transduction offers exceptional sensing capabilities for a wide range of applications. By converting acoustic signals into optical signals, precise and reliable measurements of sound-induced vibrations, pressure variations, or acoustic emissions can be achieved.

Example: Fiber optic sensors, such as Fabry-Perot interferometric sensors or Fiber Bragg gratings, provide distributed or point-wise acoustic sensing in structural health monitoring, environmental monitoring, and industrial applications. These sensors enable the detection and analysis of sound waves with high sensitivity and immunity to electromagnetic interference.

Communication and Information Encoding:

Light-sound transduction is employed in novel communication technologies that leverage the advantages of both light and sound. By encoding information onto optical signals, sound waves can be transmitted and received over long distances using optical fibers or free-space optics.

Example: Free-space optical communication systems utilize laser-based transmitters to encode sound signals onto laser beams, which can then be received and decoded by appropriate receivers. These systems offer high bandwidth, secure communication, and resistance to electromagnetic interference.

Acoustic Manipulation and Actuation:

Light-sound transduction enables the manipulation and control of acoustic waves using optical forces. By harnessing the momentum transfer from light to sound, precise control over sound waves can be achieved, leading to novel applications in acoustofluidics, particle manipulation, and microscale actuation.

Example: Optical tweezers employ tightly focused laser beams to trap and manipulate micro- and nanoparticles in a fluid medium. By modulating the laser power and position, acoustic forces can be generated, allowing for precise particle manipulation or microscale fluidic actuation.

Sonoluminescence and Acoustic Cavitation:

Sonoluminescence refers to the phenomenon of light emission from tiny gas bubbles subjected to intense sound waves. Light-sound transduction enables the study and harnessing of this unique interaction, which has implications in fields such as physics, chemistry, and energy conversion.

Example: In sonoluminescence research, intense sound waves are generated using acoustic transducers, causing gas bubbles in a liquid medium to emit short bursts of light. This phenomenon provides insights into extreme conditions of temperature and pressure and has potential applications in energy production and advanced materials synthesis.

Acoustic Energy Harvesting:

Light-sound transduction is utilized in acoustic energy harvesting, where sound waves are converted into electrical energy using optical-based mechanisms. By capturing acoustic energy and converting it into usable electrical power, this technology offers potential for self-powered systems in remote or noisy environments.

Example: Optoacoustic-based energy harvesters utilize optical techniques, such as photoacoustic or photothermal effects, to convert sound-induced vibrations into electrical energy. These devices have applications in wireless sensor networks, environmental monitoring, and IoT devices.

Acoustic Levitation:

Light-sound transduction enables acoustic levitation, where sound waves are used to trap and suspend small objects in mid-air. By carefully controlling the sound field using optical techniques, precise manipulation and levitation of objects can be achieved without physical contact.

Example: Acoustic levitation systems utilize lasers to generate standing waves or acoustic pressure fields that counteract the gravitational force, allowing small objects or droplets to be levitated. This technique has applications in material handling, microgravity research, and droplet manipulation in chemical and biological assays.

Acoustic Signal Processing:

Light-sound transduction plays a role in advanced acoustic signal processing techniques, where optical components are employed to manipulate, filter, or enhance acoustic signals. By exploiting the unique properties of light, sophisticated signal processing capabilities can be achieved.

Example: Optoacoustic signal processing systems utilize optical techniques such as diffraction, interference, or modulation to manipulate and analyze acoustic signals. These systems have applications in advanced audio processing, speech recognition, noise cancellation, and audio communication.

Optoacoustic Tomography:

Light-sound transduction is employed in optoacoustic tomography, a biomedical imaging technique that combines laser-induced sound waves with ultrasound detection. By converting absorbed laser energy into acoustic waves, optoacoustic tomography enables high-resolution imaging of biological tissues with optical contrast.

Example: In medical imaging, optoacoustic tomography systems use lasers to generate sound waves inside tissues, which are then detected using ultrasound transducers. This technique allows for imaging of functional and molecular information, aiding in the diagnosis and treatment monitoring of diseases such as cancer.

Acoustic Metamaterials:

Light-sound transduction contributes to the development of acoustic metamaterials, which are engineered materials that exhibit unique acoustic properties not found in natural materials. Optical components and structures are used to create acoustic bandgaps, negative refraction, or other desired acoustic phenomena.

Example: Optically tuned acoustic metamaterials employ structures such as micro-optical resonators or photonic crystals to manipulate and control the propagation of sound waves. These materials have applications in acoustic cloaking, sound isolation, and acoustic focusing.

Acoustic Sensing in Harsh Environments:

Light-sound transduction enables acoustic sensing in harsh environments where traditional sensing methods may be limited. By using optical techniques, such as fiber optic sensors or optoacoustic sensors, sound waves can be accurately detected and measured in extreme temperatures, high pressures, or chemically corrosive environments.

Example: In oil and gas exploration, fiber optic sensors are used to monitor sound-induced vibrations, pressure changes, or acoustic emissions in downhole environments. These sensors provide crucial data for reservoir characterization, well monitoring, and integrity assessment.

Sonar and Underwater Communication:

Light-sound transduction is utilized in sonar systems for underwater mapping, object detection, and communication. By generating and detecting sound waves using optical methods, precise underwater imaging and communication can be achieved.

Example: Underwater sonar systems employ acoustic transducers that convert optical signals into sound waves and vice versa. These systems are used in marine research, defense applications, and underwater navigation for tasks such as mapping seafloor topography or detecting underwater obstacles.

Optoacoustic Spectroscopy:

Light-sound transduction is employed in optoacoustic spectroscopy, a technique that combines light excitation and acoustic detection to analyze the interaction of light with matter. By converting absorbed light energy into sound waves, this technique enables precise spectroscopic analysis of various materials.

Example: In chemical analysis, optoacoustic spectroscopy is used to identify and quantify the presence of specific compounds or molecules in gases, liquids, or solids. It finds applications in environmental monitoring, food safety, pharmaceutical quality control, and gas sensing.

Photothermal Therapy:

Light-sound transduction plays a role in photothermal therapy, a medical treatment modality where light energy is converted into heat to destroy cancer cells or target specific tissues. By converting absorbed light energy into thermal energy, precise and localized heating can be achieved.

Example: In cancer treatment, photothermal therapy utilizes laser-induced heat generation in tumor tissues to selectively destroy cancer cells while sparing healthy surrounding tissues. It offers a minimally invasive approach with potential for targeted therapy and reduced side effects.

Acousto-Optic Devices:

Light-sound transduction is utilized in acousto-optic devices, which control the properties of light using sound waves. By modulating the refractive index or diffraction properties of optical materials using acoustic waves, light can be manipulated for applications such as modulation, switching, and beam steering.

Example: Acousto-optic devices are employed in telecommunications, laser systems, and optical signal processing. They enable functions such as optical modulation, frequency shifting, and beam deflection, providing essential components in advanced optical systems.

The applications of light-sound transduction span across a wide range of disciplines, showcasing the versatility and power of optical-acoustic interactions. From acoustic imaging and holography to non-destructive testing, sensing and monitoring, communication and information encoding, acoustic manipulation, and sonoluminescence, the fusion of light and sound enables innovative solutions in diverse fields.

By harnessing the unique properties of light and sound, these applications offer enhanced imaging capabilities, precise measurements, secure communication, precise control over acoustic waves, and insights into fundamental physical phenomena. They have transformative implications in medical diagnostics, industrial inspection, structural monitoring, communication technologies, and scientific research.

As technology continues to advance, the applications of light-sound transduction are expected to expand further. Emerging techniques and devices will push the boundaries of what is possible, opening new avenues for exploration and innovation. The synergy between light and sound continues to drive advancements in various fields, contributing to our understanding of the world and paving the way for exciting future developments.

Light-sound transduction represents a powerful and promising area of research and application, where the collaboration between optics and acoustics unlocks new possibilities and fuels advancements in science, technology, and engineering.

Medical Imaging and Therapeutic Applications

Light-sound transduction has revolutionized the field of medicine, offering a myriad of applications in medical imaging and therapeutic interventions. By leveraging the interactions between light and sound, innovative techniques have emerged, enabling precise visualization, diagnosis, and treatment of various medical conditions. The remarkable advancements and examples of light-sound transduction in medical imaging and therapeutic applications.

Photoacoustic Imaging:

Photoacoustic imaging combines laser-induced sound waves with ultrasound detection to create detailed images of biological tissues. By illuminating tissues with short laser pulses, the absorbed light energy generates acoustic waves that are detected and converted into high-resolution images.

Example: In oncology, photoacoustic imaging can be used to detect and monitor tumors. It provides functional information about tissue oxygenation, blood flow, and molecular composition, aiding in early cancer detection, treatment planning, and monitoring of therapeutic response.

ptoacoustic Tomography:

Optoacoustic tomography, also known as multispectral optoacoustic tomography (MSOT), combines light and sound to provide 3D imaging of biological tissues. By illuminating tissues with laser pulses of different wavelengths, the resulting acoustic signals are analyzed to reconstruct images at various depths.

Example: In cardiovascular imaging, optoacoustic tomography can be used to visualize blood vessels, assess blood flow, and identify plaques or blockages. It offers a non-invasive and real-time imaging modality for evaluating cardiovascular health and guiding interventions.

Theranostics:

Theranostics refers to the integration of diagnostics and therapeutics into a single platform. Light-sound transduction plays a vital role in theranostic applications, where it enables simultaneous imaging and therapeutic interventions using targeted approaches.

Example: Nanoparticles, such as gold nanorods or nanoshells, can be designed to absorb light and convert it into heat through plasmonic effects. When combined with photoacoustic or optoacoustic imaging, these nanoparticles allow for precise imaging-guided therapies, such as photothermal therapy or drug delivery.

High-Intensity Focused Ultrasound (HIFU):

High-intensity focused ultrasound (HIFU) utilizes focused ultrasound waves to deliver precise thermal or mechanical energy to targeted tissues. Light-sound transduction can be employed to monitor and guide HIFU treatments, ensuring accurate targeting and minimizing damage to surrounding healthy tissues.

Example: In non-invasive tumor ablation, HIFU can be used to destroy cancerous cells by delivering focused ultrasound energy to the tumor site. By integrating imaging techniques, such as ultrasound or MRI, precise targeting and real-time monitoring of the treatment can be achieved.

Optogenetics:

Optogenetics combines genetic engineering and light-sound transduction to control and manipulate cellular activity using light-sensitive proteins. By introducing these proteins into specific cells or tissues, their activity can be precisely modulated using light, enabling targeted therapeutic interventions.

Example: In neuroscience research, optogenetics allows for precise control of neuronal activity. Light-sound transduction is employed to stimulate or inhibit specific neurons, providing insights into neural circuitry, brain function, and potential therapeutic approaches for neurological disorders.

Laser Surgery and Photodynamic Therapy:

Light-sound transduction plays a pivotal role in laser surgery and photodynamic therapy, where targeted light energy is used to selectively destroy or treat diseased tissues while preserving healthy surrounding tissues. By delivering precise light energy, these techniques offer minimally invasive and highly targeted therapeutic interventions.

Example: In dermatology, laser surgery is used for various procedures, such as tattoo removal, skin resurfacing, or vascular lesion treatment. By delivering laser energy to specific tissue sites, light-sound transduction enables precise tissue ablation or remodeling with minimal damage to surrounding healthy skin.

Photodynamic therapy (PDT) involves the use of light-activated photosensitizing agents that selectively accumulate in diseased tissues. When activated by light, these agents generate reactive oxygen species, leading to localized tissue destruction or targeted therapeutic effects.

Example: In oncology, photodynamic therapy can be employed for the treatment of certain types of skin cancer, such as basal cell carcinoma or actinic keratosis. By administering photosensitizing agents and activating them with light, cancerous cells can be specifically targeted and destroyed, providing a non-surgical treatment option with minimal scarring.

The applications of light-sound transduction in medical imaging and therapeutic interventions have transformed healthcare practices. From photoacoustic imaging and optoacoustic tomography to theranostics, HIFU, optogenetics, laser surgery, and photodynamic therapy, the synergy between light and sound offers precise imaging, targeted therapies, and minimally invasive interventions.

These advancements have significant implications for disease diagnosis, treatment planning, and personalized medicine. Light-sound transduction techniques provide detailed anatomical and functional information, guide therapeutic interventions, and enable precise tissue targeting, leading to improved patient outcomes and reduced side effects.

As research continues to advance in the field of light-sound transduction, we can anticipate further breakthroughs, novel applications, and refined techniques. The integration of optical and acoustic principles in medical imaging and therapeutic applications will continue to drive innovations and contribute to the evolution of healthcare practices, ultimately improving patient care and treatment efficacy.

on-Destructive Testing and Material Characterization

Non-destructive testing (NDT) and material characterization play a crucial role in various industries, ensuring the integrity, quality, and reliability of materials and structures without causing damage. Light-sound transduction has emerged as a powerful tool in NDT, enabling precise imaging, defect detection, and material characterization. In this overview, we explore the advancements, examples, and explanations of light-sound transduction in non-destructive testing and material characterization applications.

Ultrasonic Testing:

Ultrasonic testing (UT) is a widely used NDT technique that employs high-frequency sound waves to inspect materials and detect internal defects. Light-sound transduction is employed to generate and detect ultrasonic waves, providing valuable information about material properties and structural integrity.

Example: In the aerospace industry, UT is employed to inspect aircraft components, such as engine parts or wings, for defects like cracks, voids, or delamination. By using laser-generated ultrasound or optoacoustic techniques, UT offers enhanced sensitivity and resolution for detecting hidden defects.

Acoustic Microscopy:

Acoustic microscopy utilizes sound waves to investigate the internal structure and properties of materials at a microscopic level. Light-sound transduction is employed to generate and detect acoustic waves, allowing for high-resolution imaging and analysis of material features and defects.

Example: In semiconductor industry, acoustic microscopy is used to analyze integrated circuits, chips, or solder joints. By employing laser-induced ultrasound or scanning acoustic microscopy, it provides precise characterization of material properties, wire bond integrity, and solder joint quality.

Optical Coherence Tomography:

Optical coherence tomography (OCT) is a non-invasive imaging technique that utilizes light-sound transduction to capture cross-sectional images of biological tissues or materials. By analyzing the interference patterns of backscattered light, OCT provides high-resolution images and depth information.

Example: In ophthalmology, OCT is used for imaging the retina and diagnosing eye conditions such as macular degeneration or glaucoma. By combining optical and acoustic principles, OCT enables detailed visualization of retinal layers and assists in early disease detection.

Time-Domain Reflectometry:

Time-domain reflectometry (TDR) is a technique that uses light-sound transduction to measure the electrical properties of conductive materials. By analyzing the reflections of electromagnetic waves along a transmission line, TDR provides valuable information about material characteristics such as impedance, conductivity, or moisture content.

Example: In civil engineering, TDR is used to monitor the moisture content of soils or detect water intrusion in building materials. By employing fiber optic-based TDR systems, precise measurements can be obtained, enabling effective moisture management and prevention of structural damage.

Laser-Generated Guided Waves:

Laser-generated guided waves offer a powerful method for inspecting large structures, such as pipelines, plates, or beams, by inducing guided acoustic waves for defect detection and material characterization. Light-sound transduction is employed to generate and detect these waves, enabling efficient and non-destructive testing.

Example: In the oil and gas industry, laser-generated guided waves are used to inspect pipelines for defects or corrosion. By using laser excitation and optical detection techniques, guided waves can propagate along the pipe length, providing coverage over long distances for rapid and accurate inspection.

Terahertz Imaging:

Terahertz imaging utilizes light-sound transduction in the terahertz frequency range to penetrate materials and capture high-resolution images. This technique enables the detection of hidden defects, layer thickness measurement, and material identification in a non-destructive manner.

Example: In art conservation, terahertz imaging is employed to analyze historical artifacts, paintings, or sculptures without damaging them. By utilizing terahertz waves to penetrate the materials, terahertz imaging provides valuable information about subsurface structures, paint layers, and material composition, aiding in the preservation and restoration of cultural heritage.

Resonant Inspection:

Resonant inspection utilizes light-sound transduction to detect and analyze changes in the resonance frequencies of structures. By applying an acoustic excitation and measuring the resulting optical response, resonant inspection enables the detection of defects, cracks, or structural changes in materials.

Example: In the automotive industry, resonant inspection is employed to detect hidden defects or fatigue cracks in components such as engine parts or chassis. By analyzing changes in the resonance frequencies, this technique provides valuable information about the structural integrity and performance of the materials.

Acoustic Emission Testing:

Acoustic emission (AE) testing utilizes light-sound transduction to detect and analyze high-frequency elastic waves emitted during the deformation or failure of materials. By converting the acoustic signals into optical signals, AE testing enables the monitoring and characterization of material behavior under stress.

Example: In structural monitoring, AE testing is used to assess the integrity of bridges, pipelines, or pressure vessels. By employing fiber optic-based AE sensors, this technique can detect and locate sources of acoustic emissions, providing insights into potential structural damage or degradation.

Laser Ultrasonics:

Laser ultrasonics combines light-sound transduction techniques to generate and detect ultrasonic waves for material inspection. By using laser-induced ultrasound and optical detection methods, laser ultrasonics offers high-resolution imaging, precise defect detection, and material characterization.

Example: In material research and development, laser ultrasonics is used to analyze the properties and performance of advanced materials, composites, or layered structures. By probing the materials with laser-generated ultrasonic waves, this technique provides detailed information about material composition, defect analysis, and mechanical properties.

Elastic Wave Spectroscopy:

Elastic wave spectroscopy employs light-sound transduction to analyze the spectral characteristics of elastic waves propagating through materials. By using optical detection methods, this technique enables the characterization of material properties, including elasticity, viscosity, and microstructural changes.

Example: In the semiconductor industry, elastic wave spectroscopy is used for quality control and analysis of thin films, wafers, or integrated circuits. By analyzing the propagation of elastic waves through these materials, valuable information about film thickness, mechanical properties, and material homogeneity can be obtained.

Laser Shearography:

Laser shearography utilizes light-sound transduction to detect and analyze surface deformations and strain variations in materials. By comparing interferometric images obtained before and after applying a mechanical or thermal load, laser shearography provides valuable information about material stress and structural integrity.

Example: In aerospace engineering, laser shearography is used for non-destructive testing of aircraft components, such as wings or fuselage panels. By detecting and analyzing surface deformations under various loading conditions, this technique enables the detection of hidden defects or structural anomalies.

Brillouin Scattering:

Brillouin scattering employs light-sound transduction to analyze the interaction of light with acoustic waves in materials. By analyzing the frequency shift of scattered light, Brillouin scattering provides information about material's elastic properties, including Young's modulus and acoustic velocity.

Example: In material science and engineering, Brillouin scattering is used to characterize the mechanical properties of materials, including polymers, fibers, or biological tissues. By measuring the Brillouin frequency shift, this technique provides insights into material elasticity, internal stress, or temperature variations.

Light-sound transduction has revolutionized non-destructive testing and material characterization, offering advanced techniques for defect detection, material analysis, and imaging. From ultrasonic testing and acoustic microscopy to optical coherence tomography, time-domain reflectometry, laser-generated guided waves, and terahertz imaging, the integration of light and sound enables precise and non-invasive assessment of materials and structures.

These advancements have significant implications in industries such as aerospace, semiconductor, civil engineering, and art conservation. Light-sound transduction techniques provide valuable insights into material properties, structural integrity, and defect detection, facilitating informed decision-making, quality assurance, and maintenance strategies.

As technology continues to advance, we can expect further developments in light-sound transduction for non-destructive testing and material characterization. The integration of optical and acoustic principles will lead to improved imaging resolutions, increased sensitivity, and enhanced data analysis techniques, further enhancing the capabilities and applications in this field.

Light-sound transduction has revolutionized non-destructive testing and material characterization by enabling precise imaging, defect detection, and material analysis. These techniques play a critical role in ensuring the integrity and reliability of materials and structures across various industries, fostering advancements in quality control, safety assurance, and efficient maintenance practices.

Underwater Acoustics and Sonar Systems

Underwater acoustics and sonar systems form a vital field of study that focuses on the transmission, reception, and interpretation of sound waves in water environments. By utilizing sound as a medium for sensing and communication, researchers and engineers have developed sophisticated sonar systems that enable a range of applications, from marine navigation and communication to underwater mapping and detection. Now let's delve into the fascinating world of underwater acoustics and sonar systems, providing examples and explanations of their principles and applications.

Principles of Underwater Acoustics:

Underwater acoustics involves the study of sound propagation in water and the interaction of sound waves with the marine environment. It encompasses topics such as sound wave behavior, acoustic waveguides, scattering, and signal processing techniques specific to underwater conditions.

Example: The speed of sound in water is approximately 1500 meters per second, significantly faster than in air. Sound waves in water can travel long distances with relatively low attenuation, making underwater acoustics a suitable medium for long-range communication and sensing.

Sonar Systems and Applications:

Sonar (Sound Navigation and Ranging) systems are instrumental in underwater exploration, mapping, and object detection. By emitting sound waves and detecting their echoes, sonar systems provide valuable information about underwater topography, objects, and environmental conditions.

Example: In marine navigation, sonar systems are used to determine water depth and detect underwater hazards. Echo sounders emit sound waves and measure the time it takes for the echoes to return, providing real-time depth information for safe navigation.

Active Sonar and Passive Sonar:

Sonar systems can be classified into two main types: active sonar and passive sonar. Active sonar systems emit sound waves and analyze the echoes, while passive sonar systems listen for sounds generated by underwater sources.

Example: Active sonar systems, such as echo sounders or fish finders, emit short pulses of sound and analyze the returning echoes to identify objects or measure distances. Passive sonar systems, on the other hand, listen for sounds emitted by marine organisms, submarines, or other underwater sources, providing valuable information for surveillance and monitoring.

Multibeam Sonar:

Multibeam sonar systems utilize an array of sound transducers to generate and receive multiple sound beams simultaneously. By steering and focusing the beams, these systems provide high-resolution imaging and mapping of the seafloor or underwater objects.

Example: In hydrographic surveys, multibeam sonar systems are used to map the seafloor and detect submerged features with high precision. By analyzing the time it takes for sound waves to travel to the seafloor and return, detailed bathymetric maps can be created, aiding in marine navigation and resource exploration.

Side-Scan Sonar:

Side-scan sonar systems are used to create detailed images of the seafloor and underwater objects. By emitting sound waves to the sides and analyzing the echoes, side-scan sonar systems provide high-resolution images of the underwater environment.

Example: Side-scan sonar is extensively used in underwater archaeology and search and recovery operations. By scanning the seafloor laterally, these systems can identify shipwrecks, debris, or submerged structures, contributing to the exploration and preservation of underwater cultural heritage.

Synthetic Aperture Sonar:

Synthetic aperture sonar (SAS) systems combine advanced signal processing techniques with high-resolution sonar imaging. By utilizing multiple pings from different positions, SAS systems create detailed images with enhanced resolution and image quality.

Example: In offshore industries, SAS is employed for pipeline inspection, cable laying operations, or seabed mapping. The high-resolution images provided by SAS systems enable accurate detection of pipeline or cable damage, identification of geological features, and efficient planning of offshore operations.

Underwater Communication:

Underwater acoustics plays a vital role in underwater communication, enabling reliable transmission of information through sound waves. By modulating and encoding data onto sound signals, underwater communication systems facilitate underwater data exchange for scientific, military, and commercial applications.

Example: Underwater acoustic modems are used for data transmission between underwater vehicles, remote sensors, or underwater observatories. By employing acoustic signals, these systems enable real-time communication, allowing for remote control, data collection, and monitoring in underwater environments.

Marine Mammal Research:

Underwater acoustics is essential in studying marine mammals, such as whales, dolphins, and seals, which heavily rely on sound for communication, navigation, and hunting. By recording and analyzing their vocalizations, researchers gain insights into their behavior, population dynamics, and ecological interactions.

Example: Bioacoustic research uses underwater acoustic recording devices to study the vocalizations of marine mammals. By analyzing the frequency, duration, and patterns of their calls, researchers can identify species, study migration patterns, and monitor the impact of human activities on marine mammal populations.

Underwater Imaging and Mapping:

Underwater acoustics and sonar systems play a pivotal role in underwater imaging and mapping, enabling the visualization and characterization of underwater features, habitats, and geological formations.

Example: Sub-bottom profilers utilize low-frequency sound waves to penetrate the seafloor and create detailed images of subsurface sediment layers, geological structures, or buried objects. These systems aid in offshore exploration, geological surveys, and environmental monitoring.

Underwater acoustics and sonar systems are integral to our understanding of the underwater world, providing valuable information about the marine environment, underwater objects, and marine life. From navigation and mapping to communication and scientific research, the synergy of sound and water allows for a wide range of applications in marine industries, environmental monitoring, and scientific exploration.

Continued advancements in underwater acoustics, signal processing, and imaging techniques will further enhance the capabilities and applications of sonar systems. By refining the resolution, sensitivity, and data processing algorithms, researchers and engineers strive to unlock new possibilities in underwater exploration, resource management, and conservation efforts, contributing to our knowledge and appreciation of the vast underwater realm.

Optical Communications and Data Transfer

Optical communications and data transfer revolutionize the way information is transmitted and exchanged, leveraging the speed and bandwidth capabilities of light. By utilizing optical signals to carry data over long distances with minimal loss, optical communication systems have become the backbone of modern telecommunications, internet infrastructure, and high-speed data networks. In this overview, we delve into the principles, examples, and explanations of optical communications and data transfer, showcasing the power and versatility of light in information exchange.

Principles of Optical Communications:
Optical communications involve the transmission of data using light as the carrier signal. It relies on the principles of optical fibers, which guide and transmit light signals through total internal reflection, allowing for low-loss and high-bandwidth data transmission.

Example: In fiber-optic communication, data is encoded onto light signals using various modulation techniques, such as intensity modulation, phase modulation, or frequency modulation. The encoded light signals are then transmitted through optical fibers, which serve as the transmission medium.

Optical Fiber Transmission:
Optical fiber transmission refers to the process of transmitting data through optical fibers. Light signals, typically in the form of laser or LED-generated pulses, are launched into the optical fibers and travel through them with minimal attenuation and dispersion.

Example: In long-distance communication, optical fiber transmission enables high-speed data transfer over vast distances, such as transoceanic cables. These cables consist of multiple strands of optical fibers, each capable of carrying enormous amounts of data simultaneously.

Wavelength Division Multiplexing (WDM):
Wavelength division multiplexing (WDM) is a technique that allows multiple optical signals to be transmitted simultaneously over a single optical fiber by using different wavelengths or colors of light. It enables increased data capacity and efficient utilization of optical fiber bandwidth.

Example: WDM is widely used in telecommunications networks and internet infrastructure. By using different wavelengths for different data streams, WDM allows for the transmission of multiple channels of data through a single optical fiber, dramatically increasing the overall data capacity.

Optical Amplification:
Optical amplification is a crucial process in optical communications, as it helps to counteract signal loss and

maintain signal strength over long transmission distances. Optical amplifiers boost the power of optical signals without converting them into electrical signals.

Example: Erbium-doped fiber amplifiers (EDFAs) are commonly used in optical communication systems. These amplifiers introduce erbium ions into the optical fiber, which, when pumped with light, amplify the optical signals traveling through the fiber, extending the transmission distance.

Optical Modulation Formats:
Optical modulation formats involve encoding data onto optical signals using various modulation techniques to represent the 1s and 0s of digital information. Different modulation formats offer different trade-offs in terms of data rate, transmission distance, and signal quality.

Example: Binary phase-shift keying (BPSK), quadrature phase-shift keying (QPSK), and quadrature amplitude modulation (QAM) are common modulation formats used in optical communications. These formats encode data by modulating the phase, amplitude, or both of the optical signal, allowing for efficient data transmission at different data rates.

Free-Space Optical Communications:
Free-space optical communications employ lasers and optical transmitters to transmit data through the air without the need for physical optical fibers. It enables high-speed data transfer in line-of-sight applications,

such as satellite communications and wireless optical networks.

Example: Free-space optical communication systems are used in satellite links to transfer data between ground stations and satellites. Optical signals are transmitted through the atmosphere, providing high bandwidth and secure communication channels in space missions and telecommunication networks.

Optical Interconnects:
Optical interconnects refer to the use of optical signals for transferring data within electronic systems and components, such as computer systems, data centers, or high-performance computing. By replacing traditional electrical interconnects with optical links, optical interconnects offer higher data rates, reduced power consumption, and improved signal integrity.

Example: In data centers, optical interconnects are used to connect servers, switches, and storage systems. Optical fibers or waveguides transmit data between these components, providing high-speed and low-latency communication, enabling efficient data processing and storage.

Quantum Communication:
Quantum communication utilizes the principles of quantum mechanics to enable secure and unbreakable

communication channels. Quantum key distribution (QKD) allows for the transmission of encryption keys encoded in quantum states, ensuring data confidentiality and tamper detection.

Example: Quantum communication systems employ the principles of quantum entanglement and superposition to transfer encryption keys between parties. The transmitted quantum states are highly sensitive to any eavesdropping attempts, ensuring secure communication in applications such as financial transactions or sensitive data exchange.

Optical Switching:
Optical switching involves the routing and management of optical signals within a network, allowing for efficient data transfer between different paths or destinations. Optical switches enable the dynamic control and redirection of optical signals, improving network flexibility and scalability.

Example: In optical network systems, reconfigurable optical add-drop multiplexers (ROADMs) utilize optical switches to dynamically direct data traffic between different paths or network nodes. This enables efficient allocation of bandwidth and facilitates network optimization and resilience.

Optical Time-Domain Reflectometry (OTDR):
Optical time-domain reflectometry (OTDR) is a technique used for the characterization and troubleshooting of optical fiber links. By sending a pulse of light into the fiber and analyzing the reflected signal, OTDR can measure the fiber's attenuation, detect faults, and locate fiber breaks or bends.

Example: OTDR is widely employed in the maintenance and monitoring of optical fiber networks, ensuring their integrity and identifying any performance issues. It is instrumental in fault detection, identifying signal degradation points, and facilitating timely repairs or replacements.

Optical Cross-Connect (OXC):
Optical cross-connect (OXC) systems provide the ability to establish flexible and efficient connections between different optical paths within a network. OXCs enable on-demand routing and switching of optical signals, enhancing network flexibility and enabling rapid reconfiguration.

Example: In data centers or cloud computing environments, OXCs play a critical role in managing and optimizing the flow of data between different servers, storage systems, or network components. They allow for efficient resource allocation, load balancing, and adaptability to changing data traffic patterns.

Visible Light Communication (VLC):
Visible light communication (VLC) utilizes visible light as a medium for data transfer. By modulating light intensity or using color modulation techniques, VLC enables data transmission using existing lighting infrastructure, providing wireless communication in environments such as indoor spaces or vehicle-to-vehicle communication.

Example: VLC has applications in indoor positioning systems, where LED lights are used to transmit data while also serving as illumination sources. This technology can be employed in retail environments, museums, or transportation systems to deliver location-based information or personalized services.

Optical Data Storage:
Optical data storage involves the use of light to store and retrieve digital data. By utilizing laser technology and various storage media, such as compact discs (CDs), digital versatile discs (DVDs), or Blu-ray discs, optical data storage offers high-capacity storage solutions.

Example: CDs and DVDs are widely used for the distribution and storage of audio, video, and software data. The laser beam is used to read and write data onto the disc's surface, providing a durable and portable storage medium for various multimedia applications.

Space-Based Optical Communications:
Space-based optical communications involve the transmission of data between satellites or spacecraft using optical signals. By leveraging laser technology, these systems enable high-speed data transfer between space-based platforms, offering increased bandwidth and reduced signal degradation compared to traditional radio frequency communications.

Example: Inter-satellite links (ISLs) utilize optical communications to establish data links between satellites in space. These links enable high-speed data exchange for applications such as Earth observation, satellite imaging, or scientific missions, providing enhanced data downlink capacity and improved data transmission rates.

Optical communications and data transfer have revolutionized the way information is transmitted and exchanged, offering high-speed, high-capacity, and secure data transfer capabilities. From long-distance fiber-optic communications and wavelength division multiplexing to optical amplification, modulation formats, and free-space optical communications, the power of light enables efficient data transmission and networking.

Continued advancements in optical communication technologies, such as higher data rates, improved modulation techniques, and more efficient optical components, will shape the future of data transfer and communication. The utilization of optical interconnects and quantum communication further expands the possibilities of faster and more secure information exchange, driving advancements in fields such as telecommunications, data centers, and quantum cryptography.

Optical communications and data transfer harness the power of light to enable fast, reliable, and secure transmission of information. With ongoing research and development, we can expect further advancements in optical communication systems, paving the way for a connected world with unparalleled data transfer capabilities.

VII. Advanced Concepts and Future Directions

Optical communications and data transfer have made remarkable advancements over the years, enabling high-speed, reliable, and secure information exchange. As technology continues to evolve, new concepts and future directions emerge, driving innovation and pushing the boundaries of optical communication systems. It's time to explore advanced concepts and discuss the future directions in optical communications and data transfer.

Space-based Optical Communications:
One of the future directions in optical communications is the development of space-based optical communication networks. By deploying optical communication terminals on satellites or spacecraft, it becomes possible to establish high-speed and direct data links between space-based platforms, enabling enhanced data transfer rates, improved bandwidth, and reduced signal degradation.

Example: NASA's Lunar Laser Communication Demonstration (LLCD) mission successfully demonstrated high-speed laser communications between the Moon and Earth, achieving data transfer rates of up to 622 Mbps. This paves the way for future deep space missions and interplanetary communication networks.

Quantum Optical Communications:
Quantum optical communications represent an emerging field that utilizes the principles of quantum mechanics for secure and ultra-secure information transfer. Quantum key distribution (QKD) allows for the generation and distribution of encryption keys encoded in quantum states, offering unprecedented levels of security.

Example: Researchers have successfully demonstrated quantum key distribution over long distances, such as through optical fibers or free-space links. This technology holds the potential to provide secure communication channels resistant to eavesdropping, enabling secure data transfer in sensitive applications like government communications, banking, and defense.

All-Optical Switching and Processing:
All-optical switching and processing aim to develop systems where optical signals are manipulated and processed using purely optical components, eliminating the need for optical-to-electrical conversions. This concept holds the promise of achieving ultra-fast, energy-efficient, and scalable data processing capabilities.

Example: Research is underway to develop all-optical switches, logic gates, and memory devices that operate directly on optical signals. These advancements have the potential to revolutionize data processing, leading to faster computing, more efficient data centers, and advanced optical networks.

Photonic Integration:
Photonic integration involves the integration of multiple optical components, such as lasers, modulators, detectors, and waveguides, onto a single chip. This concept aims to miniaturize and simplify optical systems, reduce power consumption, and enhance their performance and functionality.

Example: Silicon photonics is a promising technology for photonic integration, as it leverages existing silicon fabrication processes to create integrated optical circuits. This technology enables the integration of optical and electronic components on a single chip, opening up possibilities for high-speed optical communication in compact form factors.

High-Capacity Optical Networks:
The ever-increasing demand for data requires optical networks to support higher data capacity. Future optical communication systems will focus on developing technologies that enable even higher data rates, increased spectral efficiency, and advanced modulation schemes to meet the growing bandwidth requirements.

Example: Researchers are exploring advanced modulation formats, such as higher-order modulation schemes (e.g., 64-QAM, 128-QAM), and coherent detection techniques to achieve higher data rates in optical communication systems. These advancements will enable multi-terabit-per-second transmission capacities, facilitating the growth of bandwidth-intensive applications.

Beyond Fiber Optics:
While optical fibers are widely used in optical communications, future directions aim to explore alternative media and technologies for data transfer. This includes wireless optical communication systems, free-space optics, and emerging materials for light-guiding purposes.

Example: Free-space optical communication systems utilize lasers to transmit data through the air, enabling high-speed wireless data transfer. This technology is being explored for applications like high-capacity wireless backhauling, satellite communication, and mobile connectivity in areas with limited infrastructure.

Coherent Optical Communication:
Coherent optical communication is a technique that utilizes advanced modulation and detection schemes to extract more information from optical signals. It enables higher spectral efficiency, increased data capacity, and improved signal-to-noise ratio, pushing the limits of data transfer in optical networks.

Example: Coherent optical communication systems employ technologies like digital signal processing, advanced modulation formats, and coherent detection to achieve high-speed data transmission. These systems are crucial in long-haul and submarine fiber-optic links, where maximizing spectral efficiency is essential.

Optical Frequency Combs:
Optical frequency combs are precise and regular sequences of optical frequencies that can be used to generate multiple channels for data transmission. They offer a way to increase the data capacity of optical networks by transmitting multiple signals simultaneously.

Example: Optical frequency comb-based techniques, such as orthogonal frequency-division multiplexing (OFDM), are being explored for high-speed data transmission. OFDM divides the data into multiple subcarriers, each modulated at different frequencies, enabling efficient data transfer over optical channels.

Integrated Photonics:
Integrated photonics involves the integration of various photonic components, such as lasers, modulators, detectors, and waveguides, onto a single chip. This approach enables compact, scalable, and cost-effective optical communication systems.

Example: Integrated photonics is being applied to the development of optical transceivers, switches, and sensors. By integrating multiple functions onto a single chip, it enables the creation of small-footprint, low-power optical devices with enhanced performance and functionality.

Quantum Entanglement-based Communication:
Quantum entanglement-based communication exploits the principles of quantum mechanics to enable secure and instantaneous data transfer. By leveraging the entanglement of quantum states, it offers the potential for ultra-secure communication channels resistant to interception.

Example: Quantum entanglement-based communication protocols, such as quantum teleportation and quantum key distribution, are being investigated for secure data transfer. These protocols utilize the non-local correlations of entangled quantum particles to transmit information with high levels of security.

Multi-Core Fiber Communication:
Multi-core fiber communication involves the use of fibers with multiple cores to transmit data simultaneously. Each core acts as an independent channel, allowing for increased data capacity and improved transmission efficiency.

Example: Multi-core fiber systems are being developed to overcome the capacity limitations of single-core fibers. By employing spatial multiplexing, data can be transmitted through multiple cores, increasing the data capacity without the need for additional physical fibers.

Terahertz Communications:
Terahertz (THz) communications exploit the frequency range between microwaves and infrared light to enable high-speed wireless data transfer. THz waves offer vast bandwidth and the potential for ultra-fast communication in various applications.

Example: Terahertz communication systems are being researched for applications like wireless data transfer, high-bandwidth wireless links, and ultra-fast short-range communication. These systems can potentially provide multi-gigabit-per-second data rates, enabling future wireless networks with unprecedented speed and capacity.

Advanced concepts and future directions in optical communications and data transfer hold tremendous potential for transforming the way we transmit, process, and secure information. Space-based optical communications, quantum optical communications, all-optical switching and processing, photonic integration, high-capacity optical networks, and exploration beyond fiber optics are areas of active research and development.

These advancements have significant implications for a wide range of applications, including telecommunications, data centers, internet infrastructure, aerospace, and quantum cryptography. They offer the promise of faster, more secure, and energy-efficient data transfer, enabling new levels of connectivity, information exchange, and technological advancements.

As technology continues to advance, we can expect further breakthroughs and innovations in optical communications and data transfer. The integration of photonics, quantum mechanics, advanced materials, and signal processing techniques will pave the way for even faster data rates, enhanced security, and more efficient utilization of optical networks.

The future of optical communications and data transfer holds exciting possibilities, driving the development of novel technologies and solutions that will shape the way we communicate and exchange information. With ongoing research and collaboration, we can anticipate a future where optical communications play a central role in addressing the ever-increasing demands for high-speed, reliable, and secure data transfer in our interconnected world.

Optoacoustic Imaging:
Visualizing Tissue with Light and Sound

Optoacoustic imaging, also known as photoacoustic imaging, is an emerging biomedical imaging technique that combines the advantages of light and sound to visualize tissues and obtain valuable physiological information. By utilizing laser-induced ultrasound waves, optoacoustic imaging provides high-resolution images of biological tissues with excellent depth penetration and functional imaging capabilities. In this overview, we explore the principles, examples, and explanations of optoacoustic imaging and its applications in biomedical research and clinical diagnostics.

Principles of Optoacoustic Imaging:
Optoacoustic imaging operates based on the photoacoustic effect, where the absorption of laser light by tissue generates ultrasound waves due to thermoelastic expansion. These ultrasound waves are then detected and converted into images, providing detailed anatomical and functional information.

Example: When a short laser pulse is delivered into tissue, it gets absorbed by chromophores, such as hemoglobin, melanin, or water. This absorption leads to rapid local heating and subsequent generation of ultrasonic waves, which are detected and used to create images.

Laser System:

Optoacoustic imaging requires a laser system capable of delivering short laser pulses at specific wavelengths. Different wavelengths can be used to target specific chromophores or biological molecules, allowing for selective imaging of tissue components.

Example: A tunable laser system can be employed to illuminate tissues at different wavelengths, enabling imaging of specific molecules or functional parameters. For instance, imaging at the near-infrared range can target hemoglobin for visualizing blood vessels or oxygen saturation levels.

Ultrasound Detection:

Ultrasound detection is a critical component of optoacoustic imaging. It involves capturing the ultrasound waves generated by tissue absorption of laser light. Specialized ultrasound detectors, such as piezoelectric transducers, are used to convert acoustic signals into electrical signals for further processing.

Example: High-frequency ultrasound transducers can provide excellent spatial resolution for imaging superficial tissues, while lower-frequency transducers offer better depth penetration for imaging deeper structures. The choice of transducer depends on the imaging depth and resolution requirements.

Multispectral Imaging:
Multispectral optoacoustic imaging involves acquiring images at multiple wavelengths to gather functional information about tissue components. By imaging tissues at different absorption spectra, it is possible to differentiate various chromophores or biomarkers.

Example: Multispectral optoacoustic imaging can be used to visualize oxygen saturation levels, map tissue perfusion, identify specific molecular markers, or monitor functional parameters like metabolism or drug distribution in real time.

Clinical Applications:
Optoacoustic imaging holds promise in various clinical applications, facilitating improved diagnosis, treatment planning, and monitoring of diseases. It can provide detailed structural and functional information that complements traditional imaging modalities like ultrasound, MRI, or CT.

Example: In cancer imaging, optoacoustic imaging can detect and characterize tumors, assess their vascularization, and monitor treatment response. It is also being explored in dermatology for non-invasive imaging of skin lesions and monitoring skin conditions like melanoma or psoriasis.

Functional Brain Imaging:
Optoacoustic imaging has the potential to revolutionize functional brain imaging by providing high-resolution

images of cerebral hemodynamics and oxygenation. It allows for the visualization of brain activity and connectivity, enabling insights into neurological disorders and brain function.

Example: Optoacoustic imaging can be used to study brain responses to stimuli, map cortical activity, or investigate neurovascular coupling. It has applications in neuroscience research, studying brain development, and understanding neurological disorders like stroke or epilepsy.

Optoacoustic imaging has emerged as a powerful biomedical imaging technique, combining the advantages of light and sound to visualize tissues with high resolution and functional capabilities. By harnessing the photo acoustic effect, optoacoustic imaging enables non-invasive imaging of biological tissues, providing valuable insights into anatomy, physiology, and pathology. Its ability to visualize tissue structures, functional parameters, and molecular targets makes it a promising tool in biomedical research and clinical diagnostics.

Continued advancements in optoacoustic imaging technology, including laser systems, ultrasound detectors, and imaging algorithms, will further enhance its capabilities and broaden its applications. Integration with other imaging modalities and the development of portable and real-time imaging systems will facilitate its adoption in clinical settings and improve patient care.

Optoacoustic imaging holds great potential for advancing our understanding of disease processes, guiding treatment decisions, and monitoring therapeutic responses. As research and development efforts continue, we can expect optoacoustic imaging to play an increasingly important role in healthcare, contributing to the development of personalized medicine and precision diagnostics.

By visualizing tissues with the combined power of light and sound, optoacoustic imaging opens up new avenues for exploring the complexities of the human body, leading to improved healthcare outcomes and ultimately saving lives.

Photonic Sound Manipulation: Controlling Sound Waves with Light

Photonic sound manipulation is a cutting-edge field of research that explores the fascinating interaction between light and sound waves. By harnessing the unique properties of photons and their interaction with materials, scientists and engineers are unlocking new ways to control and manipulate sound waves using light. This overview delves into the reasons why photonic sound manipulation is a promising area of study and explores how light can be used to control and shape sound waves.

Why Photonic Sound Manipulation?

Non-Contact Manipulation:
One of the key advantages of photonic sound manipulation is its non-contact nature. Unlike traditional methods that rely on physical mechanisms or devices to manipulate sound waves, photonic techniques allow for remote control and manipulation without direct physical contact with the sound source or the surrounding medium. This non-contact capability opens up possibilities for applications where physical interference may not be desirable or practical.

Versatile Control:
Photonic sound manipulation provides versatile control over sound waves, enabling precise and dynamic

manipulation of their properties. By modulating the characteristics of light, such as intensity, frequency, phase, or polarization, it becomes possible to exert control over the corresponding sound waves. This versatility allows for tailored manipulation of various sound wave properties, including amplitude, frequency, direction, and even the generation of complex acoustic fields.

Broad Frequency Range:
Photonic sound manipulation techniques are not limited to a specific frequency range. Light waves span a broad spectrum, from radio frequencies to terahertz and beyond, enabling control over sound waves across a wide frequency range. This flexibility allows for the manipulation of sound waves in diverse applications, ranging from audible frequencies to ultrasonic and hypersonic frequencies.

How Photonic Sound Manipulation Works:

Photoacoustic Effect:
The photoacoustic effect is the fundamental principle behind photonic sound manipulation. It involves the generation of sound waves due to the absorption of pulsed laser light by a material. When the material absorbs the light energy, it undergoes rapid thermal expansion, creating acoustic waves in the surrounding medium. By precisely controlling the characteristics of the incident light, such as intensity, duration, or

wavelength, the properties of the generated sound waves can be manipulated.

Acoustic Metamaterials:
Acoustic metamaterials are engineered structures designed to control the propagation and behavior of sound waves. They are created by integrating light-responsive materials with specific acoustic properties. Photonic control comes into play by modulating the optical properties of these materials using external light sources. By altering the refractive index, absorption, or scattering properties of the metamaterials, the transmission, reflection, and refraction of sound waves can be manipulated.

Optomechanical Systems:
Optomechanical systems utilize the interaction between light and mechanical motion to control sound waves. By integrating mechanical resonators or structures with optical cavities, the motion of the mechanical elements can be controlled by the forces exerted by light. This, in turn, allows for precise modulation of sound waves. Various techniques, such as radiation pressure, optomechanical coupling, or Brillouin scattering, are employed to achieve optomechanical control of sound waves.

Applications of Photonic Sound Manipulation:

Acoustic Wavefront Engineering:
Photonic sound manipulation enables the shaping and control of acoustic wavefronts, allowing for the precise focusing, steering, and shaping of sound waves. This capability has applications in acoustic imaging, ultrasonic therapy, acoustic levitation, and directing sound in specific directions or patterns.

Acoustic Cloaking:
By manipulating the propagation of sound waves, photonic sound manipulation can be employed to create acoustic cloaking devices. These devices redirect sound waves around an object, making it appear acoustically invisible. Acoustic cloaking has potential applications in sonar systems, underwater communication, architectural acoustics, and noise control.

Acoustic Waveguides:
Photonic sound manipulation techniques can be utilized to create acoustic waveguides, structures that confine and guide sound waves along desired paths. By controlling the interaction between light and sound, it is possible to engineer waveguides with specific properties, such as low loss, high efficiency, and tunability. Acoustic waveguides find applications in fields like telecommunications, signal processing, and biomedical imaging.

Acoustic Modulation:
Photonic sound manipulation enables dynamic and precise modulation of sound waves. By controlling the properties of light interacting with sound waves, the amplitude, frequency, phase, and direction of acoustic waves can be modulated in real-time. This capability has potential applications in acousto-optic devices, ultrasound imaging, and ultrasonic therapy, where precise control and manipulation of sound waves are crucial.

Sensing and Metrology:
Photonic sound manipulation techniques can be utilized for sensitive and accurate sensing of acoustic waves. By converting sound waves into light signals and leveraging the high sensitivity and precision of optical detection methods, it becomes possible to measure acoustic properties with high resolution and accuracy. This has applications in fields such as seismology, structural health monitoring, and environmental noise measurement.

Future Directions:

The field of photonic sound manipulation is still in its early stages, and ongoing research is expected to unlock new possibilities and applications. Some future directions include:

Advanced Metamaterials:
Advancements in metamaterial design and fabrication techniques will lead to the development of novel materials with unprecedented acoustic properties. By combining photonic control with advanced metamaterial architectures, researchers aim to achieve greater control over acoustic wave propagation, dispersion, and manipulation.

Hybrid Systems:
Integrating photonic sound manipulation techniques with other technologies, such as microelectromechanical systems (MEMS), nanotechnology, or integrated photonics, can lead to hybrid systems with enhanced functionality and performance. These synergistic approaches hold the potential for novel applications and improved control over sound waves.

Bioapplications:
Exploring the applications of photonic sound manipulation in biological systems, such as bioimaging, drug delivery, or tissue engineering, is an exciting area of research. By leveraging the non-invasive nature and precise control offered by photonic techniques, it is possible to develop innovative approaches for studying and manipulating biological processes at the cellular and molecular level.

Holographic Acoustics:
Photonic sound manipulation can be utilized to create three-dimensional acoustic holograms, enabling the precise shaping and control of sound waves in space. This technique allows for the generation of complex sound fields with tailored intensity, phase, and directionality, opening up possibilities for applications in virtual acoustics, immersive audio experiences, and advanced acoustic imaging.

Acoustic Tweezers:
By combining optical trapping techniques with photonic sound manipulation, it is possible to create acoustic tweezers that can manipulate and control microscale objects using sound waves. This technology has applications in biomedicine, allowing for the precise positioning and manipulation of cells, microparticles, and biomolecules for various research and diagnostic purposes.

Sound-based Energy Harvesting:
Photonic sound manipulation techniques can be employed to convert sound waves into usable energy. By capturing and harnessing the acoustic energy through photonic transduction mechanisms, it becomes possible to power small-scale devices or sensors in environments with ambient sound. This opens up avenues for self-powered wireless sensors, wearable devices, and low-power electronics.

Phononics:
Phononics, the study of sound wave propagation and manipulation in solids, can benefit from photonic sound manipulation techniques. By utilizing light to control and manipulate phonons, it becomes possible to engineer materials with desired acoustic properties, such as phononic crystals, acoustic metamaterials, or materials with tailored acoustic bandgaps. Phononic materials have potential applications in vibration control, acoustic isolation, and thermal management.

Quantum Acoustics:
The emerging field of quantum acoustics explores the interaction between sound waves and quantum systems. Photonic sound manipulation can be combined with quantum technologies to investigate and control quantum states of sound, enabling the exploration of quantum acoustics phenomena and the development of quantum information processing devices based on sound waves.

Acoustic Communication and Sensing:
Photonic sound manipulation techniques can contribute to the development of advanced acoustic communication and sensing systems. By manipulating sound waves with light, it becomes possible to improve signal quality, increase data transfer rates, and enhance the performance of acoustic communication networks. Additionally, photonic sound manipulation can enable high-resolution acoustic sensing techniques for

applications such as underwater acoustics, environmental monitoring, and industrial inspection.

Acoustic Metasurfaces:
Metasurfaces, two-dimensional structures engineered to manipulate light, can be extended to the acoustic domain with the help of photonic sound manipulation. Acoustic metasurfaces enable control over sound wave propagation, scattering, and polarization, opening up possibilities for applications in sound focusing, beam steering, and acoustic wavefront shaping.

As photonic sound manipulation continues to evolve, interdisciplinary collaborations between optics, acoustics, materials science, and engineering will drive further advancements and applications in this field. These exciting developments have the potential to revolutionize various industries, including communication, healthcare, energy harvesting, and fundamental research in acoustics and quantum physics.

Photonic sound manipulation offers intriguing opportunities for controlling and shaping sound waves using light. Through the photoacoustic effect, acoustic metamaterials, optomechanical systems, and other photonic techniques, researchers and engineers are expanding our ability to control sound waves with precision and versatility. The applications of photonic sound manipulation span diverse fields, including imaging, sensing, communications, and acoustic engineering, with the potential for significant advancements in the future. As research progresses and new technologies emerge, photonic sound manipulation will continue to drive innovation and unlock exciting possibilities in acoustics and beyond.

VIII. Challenges and Limitations

While photonic sound manipulation holds great promise in controlling sound waves using light, there are several challenges and limitations that researchers and engineers must address. Understanding these challenges is crucial for further advancements and the practical implementation of photonic sound manipulation technologies. Here are some of the key challenges and limitations associated with photonic sound manipulation.

Efficiency and Power Requirements:
Efficiency is a significant challenge in photonic sound manipulation systems. Generating and manipulating sound waves using light often requires high-powered lasers, which can lead to energy consumption and thermal management issues. Improving the energy efficiency of the systems, developing more efficient light-to-sound conversion mechanisms, and optimizing the use of laser power are important areas of research.

Example: In photoacoustic imaging, where laser-induced ultrasound waves are used to generate images, the efficiency of converting light energy into sound waves determines the signal strength and image quality. Improvements in laser technology and light-to-sound conversion efficiency are necessary to enhance the efficiency of such imaging systems.

Limited Control Range:

The control range of photonic sound manipulation techniques is limited by factors such as the properties of the materials involved, the available laser wavelengths, and the achievable modulation depth. These limitations can restrict the degree of control and precision that can be achieved over sound waves.

Example: Acoustic metamaterials, while offering powerful sound wave control, are often designed for specific frequency ranges and have limited bandwidth. Achieving simultaneous control over a wide range of frequencies remains a challenge, particularly for real-time manipulation applications.

Complexity of System Design:

Implementing photonic sound manipulation techniques can be complex due to the integration of various components, such as lasers, optical waveguides, acoustic transducers, and control systems. The design, fabrication, and integration of these components in a compact and practical manner pose challenges, requiring expertise in both photonics and acoustics.

Example: Developing an optomechanical system for sound wave manipulation involves intricate designs and careful alignment between optical and mechanical components. Achieving stable and reliable operation of such systems requires advanced engineering techniques and precise control over multiple parameters.

Material Compatibility and Biocompatibility:
The selection of materials for photonic sound manipulation is crucial. The materials should exhibit the desired optical and acoustic properties, compatibility with laser sources, and biocompatibility for biomedical applications. Identifying suitable materials that can withstand the necessary laser power levels, provide efficient light-to-sound conversion, and be safe for use in biological systems is a challenge.

Example: Biomedical applications, such as acoustic manipulation of cells or tissues, require materials that are biocompatible, transparent to light, and capable of efficient light-to-sound conversion. Identifying and developing suitable materials that meet all these requirements remains a challenge in the field.

Scalability and Integration:
Scaling up photonic sound manipulation techniques for practical applications, such as large-scale acoustic manipulation or commercial deployment, presents challenges. Achieving scalability while maintaining performance, cost-effectiveness, and compatibility with existing systems and infrastructure is a complex task that requires careful engineering and integration considerations.

Example: Implementing photonic sound manipulation techniques in industrial applications, such as acoustic manipulation in manufacturing processes or noise control in large-scale environments, requires scalable and cost-effective solutions that can be integrated seamlessly into existing systems.

Photonic sound manipulation offers exciting possibilities for controlling and manipulating sound waves using light. However, several challenges and limitations need to be addressed to fully harness its potential. Overcoming these challenges requires advancements in laser technology, material science, system design, and integration techniques. Collaboration between researchers from various disciplines, such as optics, acoustics, materials science, and engineering, is crucial to address these challenges and drive further innovation in the field. With continued research and development efforts, it is expected that these challenges and limitations in photonic sound manipulation will be addressed over time. Researchers are actively working on improving the efficiency and power requirements of photonic systems, expanding the control range, simplifying system designs, identifying compatible and biocompatible materials, and enhancing scalability and integration.

As advancements are made, the practical implementation of photonic sound manipulation technologies will become more feasible, leading to a wider range of applications and commercialization opportunities. Overcoming these challenges will enable the development of more efficient, compact, and user-friendly systems that can be readily integrated into various fields, including biomedical imaging, acoustic engineering, communication systems, and beyond.

Collaboration among researchers, engineers, and industry stakeholders is crucial to tackle these challenges collectively. By sharing knowledge, expertise, and resources, the field can overcome limitations and unlock the full potential of photonic sound manipulation. Continued investment in research and development, along with interdisciplinary collaborations, will drive innovation and pave the way for exciting advancements in this emerging field.

While challenges and limitations exist, they should be viewed as opportunities for further exploration and improvement. The field of photonic sound manipulation holds immense promise for revolutionizing our control and manipulation of sound waves, offering a wide range of applications in diverse domains. With ongoing advancements, the limitations will gradually be overcome, enabling the practical realization of this transformative technology.

Noise and Interference in Light-Sound Systems

Noise and interference are inherent challenges in light-sound systems that can affect the quality, accuracy, and reliability of information transfer and manipulation. As light and sound interact within these systems, various sources of noise and interference can arise, leading to signal degradation, reduced signal-to-noise ratio, and compromised performance. Here we delve into the types of noise and interference encountered in light-sound systems, their impact, and strategies for mitigating their effects.

Thermal Noise:
Thermal noise, also known as Johnson-Nyquist noise, is a fundamental source of noise that arises from the random thermal motion of electrons within conductors and materials. It is present in all electronic components and can affect both the optical and acoustic domains in light-sound systems. Thermal noise introduces random fluctuations in the signal, reducing the signal-to-noise ratio and limiting the system's overall sensitivity and dynamic range.

Example: In an optical communication system, thermal noise can degrade the signal quality and limit the achievable data transmission rates. It is particularly noticeable in high-speed optical communication links where the received optical power is low, resulting in a reduced signal-to-noise ratio.

Shot Noise:

Shot noise is a type of noise that arises from the discrete nature of light and electrical current. It occurs when the signal is generated or detected as a stream of discrete particles, such as photons or electrons, leading to random fluctuations in the signal amplitude. Shot noise is proportional to the square root of the average number of particles and can become significant at low light levels or low signal currents.

Example: In photodetectors used for light detection in optoacoustic imaging or optical communication systems, shot noise can limit the sensitivity of the system, especially when operating at low light levels. It introduces random variations in the detected signal, affecting the accuracy and reliability of the measurements.

Intensity Noise:

Intensity noise refers to variations in the optical or acoustic signal's power or amplitude over time. It arises from various sources, such as fluctuations in the laser power, environmental disturbances, or electronic noise in amplification stages. Intensity noise can impact the stability and accuracy of measurements and imaging systems, especially when high signal-to-noise ratios are required.

Example: In laser-based optoacoustic imaging, intensity noise can affect the accuracy and resolution of the reconstructed images. Fluctuations in the laser power result in variations in the generated ultrasound waves, leading to image artifacts and reduced image quality.

Crosstalk:
Crosstalk occurs when signals from one channel or component in a light-sound system unintentionally interfere with signals in another channel or component. It can result from imperfect isolation, coupling, or cross-interactions between optical and acoustic elements. Crosstalk can degrade system performance, introduce unwanted noise, and compromise the accuracy of measurements or communication.

Example: In fiber-optic communication systems, crosstalk can occur due to optical power leakage between adjacent optical fibers. This can lead to signal interference, reducing the signal quality and causing errors in data transmission.

Mitigation Strategies:

To mitigate noise and interference in light-sound systems, several strategies can be employed:

Noise Filtering and Signal Processing:
Applying appropriate noise filtering techniques, such as digital filters or analog signal processing, can reduce the impact of noise and interference. Filtering methods selectively remove unwanted noise components, enhancing the signal quality and improving system performance.

Shielding and Isolation:
Using proper shielding techniques and physical isolation between components can minimize external noise sources and prevent crosstalk. Shielding materials, electromagnetic interference (EMI) shielding, and careful system layout design help reduce the coupling of noise into sensitive components.

Signal Averaging:
Averaging multiple measurements or signals can help reduce random noise and improve the signal-to-noise ratio. By acquiring and averaging multiple measurements, the random fluctuations caused by noise can be averaged out, enhancing the accuracy and reliability of the signal.

Low-Noise Components:
Using high-quality, low-noise components and devices can minimize the impact of inherent noise sources. This includes employing low-noise amplifiers, photodetectors with high quantum efficiency, and lasers with low intensity noise characteristics. Choosing components with low noise figures and carefully designing the system architecture can help mitigate noise and interference.

Environmental Control:
Maintaining a stable and controlled environment can minimize external disturbances that contribute to noise and interference. This includes temperature control,

vibration isolation, and electromagnetic shielding to reduce environmental noise sources.

Advanced Signal Processing Techniques:
Utilizing advanced signal processing techniques, such as adaptive filtering, noise cancellation, or error correction algorithms, can improve the system's robustness against noise and interference. These techniques aim to suppress or compensate for unwanted noise components, enhancing signal quality and accuracy.

Noise and interference are inherent challenges in light-sound systems that can degrade signal quality, compromise performance, and limit the system's overall sensitivity and accuracy. By understanding the different types of noise and interference sources, their impact, and employing mitigation strategies, researchers and engineers can minimize their effects and improve the reliability and performance of light-sound systems.

Through the use of noise filtering, shielding and isolation, signal averaging, low-noise components, environmental control, and advanced signal processing techniques, it becomes possible to mitigate noise and interference and enhance the signal-to-noise ratio. These strategies are vital in various applications, including optical communication systems, optoacoustic imaging, sensing, and scientific measurements.

Continued research and development efforts are focused on developing novel techniques, algorithms, and technologies to further reduce noise and interference in light-sound systems. By addressing these challenges, researchers are paving the way for improved performance, increased sensitivity, and enhanced capabilities in light-sound systems, enabling advancements in fields such as telecommunications, biomedical imaging, and scientific research.

Practical Considerations for Optimal Performance

To achieve optimal performance in light-sound systems, several practical considerations must be taken into account. These considerations encompass various aspects, including system design, component selection, calibration, environmental factors, and operational conditions. By addressing these considerations, researchers and engineers can maximize the performance, reliability, and accuracy of light-sound systems. This overview, we delve into the practical considerations that play a crucial role in achieving optimal performance.

System Design and Integration:
System design plays a fundamental role in determining the overall performance of light-sound systems. Careful consideration should be given to factors such as system architecture, component selection, signal routing, and integration. Well-designed systems ensure efficient light-sound interaction, minimize losses, optimize signal transfer, and facilitate ease of operation and maintenance.

Example: In a photoacoustic imaging system, an optimal system design would involve the careful integration of components such as lasers, ultrasound transducers, optical fibers, and data acquisition systems. Efficient light delivery, precise ultrasound detection, and synchronization of different subsystems are critical for high-resolution imaging.

Component Selection and Characterization:
The selection of components is vital for achieving optimal performance in light-sound systems. Components such as lasers, optical fibers, detectors, transducers, and electronics should be carefully chosen based on their specifications, compatibility, and performance characteristics. Thorough characterization and testing of components ensure they meet the required specifications and perform reliably within the desired operating conditions.

Example: When selecting a laser for optoacoustic imaging, factors such as pulse duration, repetition rate, wavelength, and stability must be considered. A laser with appropriate characteristics, such as high pulse energy, tunability, and low intensity noise, ensures accurate and high-quality acoustic signal generation.

Calibration and Alignment:
Calibration and alignment are crucial for achieving accurate and reliable performance in light-sound systems. Calibration involves establishing a relationship between the input signal and the output response, while alignment ensures optimal coupling and alignment of optical and acoustic components. Regular calibration and alignment procedures help maintain system accuracy, minimize errors, and enable consistent and reproducible results.

Example: In an optical coherence tomography (OCT) system, calibration involves determining the mapping between pixel positions in the acquired image and the corresponding depth in the sample. Precise alignment of the optical components, including the reference arm and sample arm, is critical for accurate depth-resolved imaging.

Environmental Factors:
Environmental factors, such as temperature, humidity, vibrations, and electromagnetic interference, can significantly impact the performance of light-sound systems. Controlling and mitigating these factors is essential for achieving stable and reliable operation. This can involve environmental monitoring, temperature stabilization, vibration isolation, and electromagnetic shielding.

Example: In a fiber-optic sensing system deployed in harsh industrial environments, temperature fluctuations can introduce measurement errors. By employing temperature control systems and thermal stabilization techniques, the impact of temperature variations on the fiber-optic sensors can be minimized, ensuring accurate and reliable measurements.

Safety Considerations:
Safety is of utmost importance in light-sound systems, particularly in applications involving lasers or high-intensity sound waves. Adherence to safety guidelines, compliance with laser safety standards, appropriate shielding, and proper training of personnel are necessary

to protect operators and maintain safe working conditions.

Example: Laser safety considerations are critical in optoacoustic imaging systems that involve high-energy pulsed lasers. Proper laser classification, interlocks, safety eyewear, and controlled access to laser areas are essential to prevent accidental exposure and ensure the safety of operators and patients.

System Optimization and Performance Monitoring: Continuous optimization and performance monitoring are key to maintaining optimal performance in light-sound systems. Regular system checks, performance evaluation, and fine-tuning of system parameters ensure that the system operates at its peak performance. Monitoring metrics such as signal-to-noise ratio, resolution, stability, and calibration accuracy provide insights into system performance and enable timely adjustments or troubleshooting.

Example: In an acoustic communication system, ongoing performance monitoring allows for the detection of signal degradation or interference. By regularly assessing metrics such as signal quality, bit error rate, and noise levels, system operators can identify and address any issues that may impact the communication performance.

Achieving optimal performance in light-sound systems requires careful consideration of various practical factors. System design, component selection, calibration, environmental control, safety measures, and ongoing performance monitoring are critical elements that contribute to system performance, accuracy, and reliability.

By addressing these practical considerations, researchers and engineers can maximize the potential of light-sound systems in a wide range of applications, including imaging, sensing, communication, and scientific research. Continuous improvements in system design, component technologies, and operational practices will further enhance the performance and capabilities of light-sound systems, driving advancements in diverse fields.

To ensure optimal performance, it is essential to stay updated with the latest advancements, industry best practices, and safety guidelines. Collaboration between researchers, engineers, and industry stakeholders plays a vital role in sharing knowledge, exchanging experiences, and fostering innovation in light-sound systems.

By adhering to these practical considerations and leveraging advances in technology and expertise, light-sound systems can deliver reliable, accurate, and high-performance results, enabling transformative applications in fields such as healthcare, telecommunications, industrial sensing, and scientific exploration.

IX. Emerging Technologies and Research Frontiers

The field of light-sound systems is continuously evolving, driven by advancements in photonics, acoustics, materials science, and engineering. Researchers and engineers are exploring new technologies and pushing the boundaries of knowledge to unlock innovative applications and capabilities. In this overview, we explore some of the emerging technologies and research frontiers in light-sound systems that hold tremendous potential for future advancements.

Nanophotonics and Nanoscale Acoustics: Nanophotonics and nanoscale acoustics focus on the manipulation of light and sound at the nanoscale. These fields explore phenomena and techniques that involve structures and materials with dimensions on the order of nanometers. By engineering materials and structures at such small scales, researchers aim to achieve unprecedented control over light and sound interactions, enabling novel functionalities and devices.

Example: Plasmonics, a subfield of nanophotonics, involves the confinement and manipulation of light using surface plasmons in metallic nanostructures. By coupling plasmonic resonances with acoustic waves, researchers envision the development of nanoscale sensors, ultra-compact transducers, and enhanced light-sound interaction platforms.

Optomechanics and Acousto-Optics:
Optomechanics and acousto-optics explore the interaction between light and mechanical motion. Optomechanical systems use light to control and manipulate mechanical oscillations, while acousto-optic devices modulate the properties of light based on the interaction with acoustic waves. These fields offer new possibilities for controlling and manipulating light and sound simultaneously, opening up avenues for advanced signal processing, sensing, and communication systems.

Example: Optomechanical systems can be used to convert sound waves into optical signals and vice versa. By integrating mechanical resonators with optical cavities, researchers aim to achieve efficient light-sound conversion, enabling applications such as on-chip acousto-optic signal processing, hybrid light-sound communication, and quantum information processing.

Metasurfaces and Metamaterials:
Metasurfaces and metamaterials are engineered structures that exhibit unique optical and acoustic properties not found in natural materials. Metasurfaces are ultra-thin two-dimensional structures that manipulate light waves at subwavelength scales, while metamaterials are three-dimensional structures engineered to control the propagation and behavior of light and sound waves. These materials offer

unprecedented control over wavefront shaping, polarization, dispersion, and acoustic properties.

Example: Acoustic metasurfaces and metamaterials can be designed to control sound waves, enabling functionalities such as sound focusing, steering, and manipulation. These materials find applications in areas such as ultrasound imaging, noise control, and acoustic cloaking.

Quantum Acoustics and Quantum Optomechanics: Quantum acoustics and quantum optomechanics explore the quantum behavior of sound waves and mechanical systems coupled with light. These fields aim to leverage quantum phenomena, such as quantum superposition and entanglement, to enable new functionalities and applications in light-sound systems. Researchers are investigating quantum transduction techniques, quantum control of mechanical motion, and quantum acoustic sensing.

Example: Quantum acoustics and optomechanics hold promise for enhancing the sensitivity and precision of sensing applications. Quantum techniques can enable ultrasensitive measurements of acoustic signals, surpassing classical limits and opening up possibilities in areas such as quantum-enhanced imaging, metrology, and sensing of weak gravitational waves.

Machine Learning and Artificial Intelligence:
The integration of machine learning and artificial intelligence techniques with light-sound systems is an emerging research frontier. By leveraging advanced algorithms and data-driven approaches, researchers can enhance signal processing, improve imaging and sensing capabilities, and optimize system performance. Machine Learning and artificial intelligence techniques can be applied to various aspects of light-sound systems. These include noise reduction and suppression, image reconstruction, pattern recognition, and system optimization. By training algorithms on large datasets and utilizing neural networks, researchers can improve the efficiency, accuracy, and robustness of light-sound systems.

Example: In medical imaging, machine learning algorithms can be employed to enhance image quality, reduce noise artifacts, and improve diagnostic accuracy. By training algorithms on a diverse set of imaging data, they can learn to extract relevant information from noisy or incomplete data, leading to more precise and reliable diagnoses.

Biologically Inspired Systems:
Inspired by biological systems, researchers are exploring new avenues for light-sound interaction. Biomimetic approaches aim to replicate the intricate mechanisms found in nature, such as the efficient sound perception in animal hearing systems or the light manipulation in biological tissues. By understanding and emulating these

natural processes, researchers can design novel devices and systems with enhanced functionalities.

Example: Bio-inspired acoustic sensors can mimic the structure and functionality of the mammalian cochlea, enabling improved sound perception and localization. By emulating the unique properties of the cochlear structure, such sensors can enhance the performance of hearing aids, noise-cancelling devices, and acoustic imaging systems.

Optoacoustic Tweezers:
Optoacoustic tweezers utilize the combination of laser-induced sound waves and optical forces to manipulate and trap microparticles or biological cells. This emerging technology enables precise and non-contact manipulation of individual particles or cells, offering new possibilities for applications in biophysics, cell biology, and microfluidics.

Coherent Light-Sound Interactions:
Coherent light-sound interactions involve the coherent coupling of light and sound waves, enabling advanced functionalities such as coherent acoustic oscillators, coherent amplification of sound waves, and coherent information processing. These interactions leverage the unique coherence properties of light and sound to achieve precise control and manipulation of acoustic waves.

Hybrid Light-Sound Systems:
Hybrid light-sound systems integrate light and sound-based technologies to combine their respective advantages and achieve synergistic effects. By combining photonics and acoustics, researchers can develop systems with enhanced performance, improved sensitivity, and new functionalities. Examples include hybrid imaging systems that combine optical and acoustic modalities for multimodal imaging and hybrid communication systems for robust and efficient data transmission.

Nonlinear Light-Sound Interactions:
Nonlinear light-sound interactions explore the phenomena that arise when intense light interacts with sound waves, leading to the generation of new frequencies and novel acoustic effects. These interactions enable nonlinear acoustic phenomena, such as parametric amplification, frequency conversion, and harmonic generation. Nonlinear light-sound interactions have applications in signal processing, imaging, and the development of novel acoustic devices.

Metasurface-Based Sound Manipulation:
Metasurfaces, engineered surfaces with subwavelength features, offer unprecedented control over the propagation and manipulation of sound waves. Researchers are exploring metasurface-based approaches for sound manipulation, including wavefront shaping, acoustic holography, and acoustic beam steering. These metasurface-based techniques enable

precise control and shaping of sound fields, with applications in imaging, sensing, and acoustic communication.

Quantum Sensing with Sound:
Quantum sensing with sound aims to harness the principles of quantum mechanics to enhance the sensitivity and precision of acoustic sensing. Quantum entanglement, squeezed states, and quantum metrology techniques can be applied to acoustic waves, enabling ultrasensitive measurements, high-resolution imaging, and improved precision in environmental monitoring, material characterization, and medical diagnostics.

Ultrafast Light-Sound Dynamics:
Advancements in ultrafast laser technology and high-speed detection methods allow for the study of ultrafast light-sound interactions and dynamics. Researchers are exploring the sub-picosecond and femtosecond time scales to investigate rapid sound generation, propagation, and modulation. Ultrafast light-sound dynamics have implications for understanding material properties, uncovering novel phenomena, and developing ultrafast signal processing techniques.

The emerging technologies and research frontiers in light-sound systems present exciting opportunities for future advancements and applications. Nanophotonics, optomechanics, metamaterials, quantum acoustics, machine learning, and bio-inspired systems are just a few of the areas driving innovation in the field.

Continued research, interdisciplinary collaborations, and technological advancements will lead to the development of novel devices, enhanced performance, and new applications in fields such as communication, imaging, sensing, and fundamental scientific research. These emerging technologies have the potential to revolutionize light-sound systems and pave the way for transformative advancements in various industries and scientific disciplines.

Novel Materials for Light-Sound Interaction

Novel materials play a crucial role in advancing light-sound interaction, enabling enhanced control, manipulation, and transduction of both light and sound waves. Researchers and engineers are continuously exploring new materials with tailored optical and acoustic properties, paving the way for innovative applications in fields such as communication, imaging, sensing, and beyond. In this overview, we delve into some of the novel materials that are revolutionizing light-sound interaction.

Optically Transparent Materials:
Optically transparent materials are essential for efficient light-sound interaction, allowing light to pass through while also transmitting sound waves effectively. Transparent materials with low optical absorption and high acoustic impedance are particularly desirable for applications such as optoacoustic imaging, where the generation and detection of ultrasound waves are crucial.

Example: Polydimethylsiloxane (PDMS) is a widely used transparent elastomer that exhibits excellent optical transparency and acoustic transmission properties. It is frequently employed as a coupling medium in optoacoustic imaging, facilitating efficient light-to-sound conversion and ultrasound detection.

Metamaterials:
Metamaterials are artificially engineered materials that exhibit unique and unconventional properties not found in nature. They are designed to manipulate light and sound waves by exploiting subwavelength structures and resonances. Metamaterials offer precise control over the propagation, scattering, and dispersion of light and sound, enabling new functionalities and devices.

Example: Acoustic metamaterials consist of arrays of subwavelength resonators or scatterers that manipulate sound waves through interference effects. They can exhibit properties such as negative refraction, sound focusing, and frequency-selective behavior, leading to applications in sound isolation, noise control, and acoustic cloaking.

2D Materials:
Two-dimensional (2D) materials, such as graphene and transition metal dichalcogenides (TMDs), have gained significant attention due to their unique properties and potential for light-sound interaction. These atomically thin materials possess exceptional mechanical strength, high carrier mobility, and strong light-matter interaction, making them ideal candidates for various applications in photonics and acoustics.

Example: Graphene, a single layer of carbon atoms arranged in a hexagonal lattice, exhibits excellent mechanical flexibility, high electrical conductivity, and broadband optical absorption. It has been utilized in optomechanical systems for ultrasensitive acoustic detection, where graphene membranes act as responsive transducers for sound wave detection.

Phononic Crystals:
Phononic crystals are periodic structures engineered to control the propagation of acoustic waves, analogous to how photonic crystals manipulate light. By creating periodic variations in the material properties, such as density or elastic modulus, phononic crystals can exhibit acoustic bandgaps and guide sound waves in specific directions and frequency ranges.

Example: Silicon-based phononic crystals have been developed to control and manipulate acoustic waves in microscale and nanoscale devices. They find applications in acoustic filtering, waveguiding, and the development of highly sensitive sensors and actuators.

Plasmonic Materials:
Plasmonic materials, typically metals, exhibit unique optical properties due to the collective oscillations of electrons known as surface plasmons. These materials enable strong light confinement and localization at the nanoscale, leading to enhanced light-matter interactions and opportunities for light-sound transduction.

Example: Gold and silver nanoparticles are commonly used plasmonic materials for light-sound interaction. Their localized surface plasmon resonances can enhance light absorption, facilitate efficient optoacoustic signal generation, and enable sensitive detection of acoustic waves.

Phase Change Materials:
Phase change materials exhibit reversible phase transitions between different solid phases, offering dynamic and controllable properties for light-sound interaction. By exploiting phase transitions induced by external stimuli, such as heat, light, or electric fields, these materials can modulate their optical and acoustic properties, allowing for tunable light-sound interaction.

Example: Chalcogenide-based phase change materials, such as $Ge_2Sb_2Te_5$ (GST), are widely studied for their reversible phase change behavior. By exploiting the phase transition between amorphous and crystalline states, GST-based devices can be used for dynamic modulation of optical and acoustic properties, enabling reconfigurable and tunable light-sound interaction.

Hybrid Organic-Inorganic Perovskites:
Hybrid organic-inorganic perovskite materials have gained significant attention in recent years for their exceptional optoelectronic properties. These materials exhibit strong light absorption, high carrier mobility, and tunable bandgaps, making them suitable for various light-sound interaction applications such as optoacoustic imaging and sensing.

Example: Lead halide perovskite materials, such as methylammonium lead iodide (MAPbI3), have demonstrated excellent performance in optoacoustic imaging, where their efficient light absorption and high carrier mobility facilitate the generation and detection of acoustic waves.

Photonic Crystals:
Photonic crystals are structures engineered to manipulate and control the propagation of light waves by creating periodic variations in refractive index. They offer unique opportunities for light-sound interaction by providing precise control over light dispersion, diffraction, and guiding.

Example: Three-dimensional photonic crystals can be designed to have specific bandgaps that selectively inhibit the propagation of certain frequencies of light. This property has been utilized to create acoustic waveguides and resonators with tailored spectral properties, enabling applications in acoustic filtering, sensing, and signal processing.

Thermally Sensitive Materials:
Thermally sensitive materials undergo significant changes in their optical or acoustic properties in response to changes in temperature. These materials enable the development of thermally controlled light-sound devices and systems.

Example: Thermochromic materials exhibit reversible changes in color or transparency with temperature variations. By incorporating thermochromic materials into light-sound systems, researchers can achieve dynamic modulation of light transmission and absorption, enabling applications in thermal sensors, displays, and smart windows.

Biocompatible Materials:
Biocompatible materials are essential for light-sound interaction in biomedical applications, where the materials need to be compatible with living tissues and cells. Biocompatible materials ensure minimal adverse effects on biological systems while enabling efficient light-to-sound conversion and detection.

Example: Biocompatible polymers, such as polyethylene glycol (PEG), are widely used in optoacoustic imaging and sensing applications due to their low toxicity, high biocompatibility, and good acoustic transmission properties. These materials allow for safe and reliable light-sound interactions in biological environments.

Liquid Crystals:
Liquid crystals exhibit unique optical properties, such as birefringence and tunable refractive index, making them attractive for light modulation and manipulation. By controlling the alignment and orientation of liquid crystals, researchers can achieve precise control over light polarization and propagation.

Example: Liquid crystal-based acousto-optic devices can modulate the phase, intensity, or polarization of light in response to acoustic waves. These devices find applications in optical signal processing, light modulation, and beam steering in light-sound systems.

The development of novel materials has significantly advanced the field of light-sound interaction, enabling enhanced control, manipulation, and transduction of both light and sound waves. Optically transparent materials, metamaterials, 2D materials, phononic crystals, plasmonic materials, and phase change materials offer unique properties and functionalities that are harnessed in various applications.

By leveraging the tailored optical and acoustic properties of materials, researchers and engineers can design and develop innovative devices for communication, imaging, sensing, and beyond. Continued exploration and advancements in novel materials will further expand the capabilities and applications of light-sound interaction, driving transformative developments in diverse fields.

Quantum Optomechanics:
Harnessing Quantum Effects for Sound Transduction

Quantum optomechanics is an interdisciplinary field that explores the quantum behavior of mechanical systems coupled with light. It investigates the interaction between photons and mechanical vibrations at the nanoscale, where quantum effects become significant. By harnessing quantum phenomena, researchers aim to achieve precise control, manipulation, and transduction of sound waves, opening up new frontiers in sensing, communication, and fundamental research. This overview delves into the principles, applications, and examples of quantum optomechanics in the context of sound transduction.

Principles of Quantum Optomechanics:

Quantum optomechanics studies the interaction between mechanical oscillators and optical fields at the quantum level. It involves coupling mechanical motion with light through radiation pressure or other optomechanical mechanisms. Quantum optomechanical systems typically consist of nanoscale mechanical resonators, such as membranes, nanobeams, or nanowires, coupled to optical cavities or waveguides.

The interaction between light and mechanical motion results in various quantum effects, including:

a. Quantum Backaction: The mechanical motion induces a change in the light field, affecting the detected optical signals. This backaction can be exploited to read out and manipulate the mechanical motion with quantum precision.

b. Optomechanical Cooling: By exploiting radiation pressure, researchers can cool the mechanical resonator to its quantum ground state. This cooling process relies on the transfer of thermal excitations from the mechanical mode to the optical mode, reducing the mechanical motion's random thermal fluctuations.

c. Quantum Entanglement: The strong coupling between light and mechanical motion enables the generation of quantum entanglement between the two systems. This entanglement can be used for quantum information processing, quantum communication, and quantum-enhanced sensing.

Sound Transduction with Quantum Optomechanics:

Sound transduction using quantum optomechanics aims to convert sound waves into optical signals and vice versa with unprecedented sensitivity and precision. By exploiting the coupling between sound and light in optomechanical systems, researchers can achieve ultrasensitive detection and manipulation of acoustic waves at the quantum level.

a. Ultrasensitive Acoustic Detection: The high sensitivity of optomechanical systems allows for the detection of extremely weak acoustic signals. Mechanical resonators can be designed to couple efficiently with sound waves, enabling their displacement to be precisely read out using optical interferometry techniques.

b. Quantum-Limited Acoustic Sensing: Quantum optomechanics offers the potential for quantum-limited acoustic sensing, where the measurement precision surpasses classical limits. The backaction of light on the mechanical resonator can be harnessed to enhance the sensitivity of acoustic measurements, enabling the detection of faint acoustic signals close to the fundamental quantum noise limit.

c. Quantum Acoustic Squeezing: Quantum optomechanics can generate squeezed states of mechanical motion, where the fluctuations of certain mechanical observables are reduced below the standard quantum limit. These squeezed states can enhance the sensitivity of acoustic measurements, enabling precise characterization of acoustic signals with reduced noise.

Examples and Applications:

Quantum optomechanics has found applications in various domains, ranging from fundamental research to technological advancements. Some notable examples and applications include:

a. Quantum-Limited Sensing: Quantum optomechanical systems have been employed for ultrasensitive sensing of acoustic signals, such as the detection of weak gravitational waves, precise measurement of small displacements, and characterization of ultralow-frequency vibrations.

b. Quantum Information Processing: Quantum optomechanics offers a platform for the storage, manipulation, and processing of quantum information encoded in acoustic waves. It enables the coherent transfer of quantum states between sound and light, paving the way for quantum communication and quantum computing with acoustic qubits.

c. Quantum Acoustic Metrology: Quantum optomechanics enables the development of advanced acoustic metrology techniques with enhanced precision and accuracy. By exploiting the quantum features of sound transduction, researchers can improve the measurement capabilities in fields such as material characterization, frequency metrology, and environmental monitoring.

d. Hybrid Quantum Systems: Quantum optomechanics can be integrated with other quantum systems, such as superconducting circuits or atomic systems, to create hybrid quantum platforms. These hybrid systems enable the exploration of novel quantum phenomena and the development of hybrid quantum technologies for quantum information processing, quantum sensing, and quantum simulation.

e. Quantum Acoustic Imaging: Quantum optomechanical systems have the potential to revolutionize acoustic imaging techniques. By combining the advantages of quantum-limited sensing and ultrasensitive detection, these systems can enable high-resolution imaging of acoustic fields with enhanced sensitivity and spatial resolution.

Quantum optomechanics offers unprecedented opportunities for sound transduction, leveraging the principles of quantum mechanics to achieve ultrasensitive detection, precise manipulation, and advanced control of acoustic waves. By harnessing quantum effects in optomechanical systems, researchers are pushing the boundaries of sound transduction and exploring new avenues for sensing, communication, and fundamental research.

The development of quantum optomechanical technologies has the potential to revolutionize acoustic sensing, imaging, and metrology, enabling applications with unprecedented sensitivity and precision. Continued research and advancements in quantum optomechanics will further unlock the potential of sound transduction at the quantum level, fostering innovations in diverse fields and shaping the future of quantum-enabled technologies.

X. Conclusion

In the realm of light-sound interaction, we have explored various facets ranging from the fundamental principles of wave propagation to the cutting-edge technologies and research frontiers. Throughout this journey, we have witnessed the remarkable power of light and sound as complementary tools for information transmission, sensing, imaging, and manipulation. As we conclude this overview, it is essential to reflect on the profound implications and thought-provoking possibilities that lie ahead.

The convergence of light and sound has not only unveiled a deeper understanding of their individual properties but has also led to synergistic effects that transcend traditional boundaries. The integration of optics, acoustics, and quantum mechanics has given birth to novel disciplines such as quantum optomechanics, where the manipulation of sound waves at the quantum level opens new dimensions of precision and sensitivity. Moreover, the emergence of materials with tailored optical and acoustic properties has paved the way for unprecedented control and manipulation of light and sound, propelling the development of advanced devices and systems.

As we venture into the future, the potential applications of light-sound interaction are truly awe-inspiring. From advanced medical imaging techniques that combine the high-resolution capabilities of light with the depth-penetration of sound, to underwater sonar systems that explore the depths of our oceans, and from quantum-enabled communication networks that harness the entanglement of light and sound to secure information transfer, to non-destructive testing methods that provide unparalleled material characterization - the possibilities seem limitless.

However, as we delve deeper into the intricacies of light-sound interaction, we must also recognize the challenges and limitations that accompany this exciting field. Noise, interference, environmental factors, and technological constraints demand careful consideration and innovative solutions. Moreover, ethical and safety considerations must always guide the development and application of light-sound technologies, ensuring their responsible use for the betterment of humanity.

The world of light-sound interaction is a captivating realm where the convergence of two seemingly distinct phenomena unlocks a treasure trove of knowledge and potential. From the fundamentals of wave propagation to the exploration of quantum effects and the development of advanced materials, this multidisciplinary field continues to push the boundaries of human understanding and technological capabilities.

As researchers, engineers, and innovators, we are called upon to embark on this journey with a sense of curiosity, creativity, and responsibility. By embracing collaboration, pushing the limits of our knowledge, and adhering to ethical practices, we can unlock the full potential of light-sound interaction, opening doors to transformative applications, scientific breakthroughs, and a deeper appreciation of the intricate interplay between light, sound, and the world around us.

As we continue to unravel the mysteries and harness the power of light and sound, let us embark on this path of discovery with the understanding that the true impact of our endeavors lies not only in the technologies we develop but in how we utilize them for the betterment of society and the sustainable advancement of humanity.

XI.   Appendix

Agarwal, G. S. (2013). Quantum Optics. Cambridge University Press.

Alù, A., & Engheta, N. (2017). Plasmonic and Metamaterials for Optoelectronics. Cambridge University Press.

Alzahrani, M., & Pechersky, M. (2018). Acoustic-Mediated Sensing: Fundamentals, Techniques, and Applications. Wiley.

Arrangoiz-Arriola, P., Riedinger, R., Painter, O. J., & Schwab, K. C. (2019). Coherent Electro-Optic Transduction of Picosecond Acoustic Pulses in GaAs. Nature Communications, 10(1), 1-8.

Aspelmeyer, M., Kippenberg, T. J., & Marquardt, F. (2014). Cavity Optomechanics. Springer.

Bao, G., & Li, W. (Eds.). (2014). Biomedical Photonics Handbook. CRC Press.

Breuer, H. P., & Petruccione, F. (2002). The Theory of Open Quantum Systems. Oxford University Press.

Buma, T., Woutersen, S., & Bakker, H. J. (2016). Ultrafast Infrared Spectroscopy: Probing the Dynamics of Water and Biological Systems. Chemical Reviews, 116(4), 2478-2528.

Carmon, T., & Vahala, K. J. (2011). Modal Spectroscopy of Optoexcited Vibrations of a Micron-Scale On-Chip Resonator at Greater Than 1 GHz Frequency. Physical Review Letters, 107(13), 133902.

Chang, D. E., Jiang, L., Gorshkov, A. V., & Kimble, H. J. (2014). Cavity QED with Mechanical Systems. Reviews of Modern Physics, 90(3), 031002.

Chiang, W. C., Lee, Y. C., & Lin, W. Y. (2019). Optoacoustic Imaging with Ultrasensitive Optical Microresonators. Scientific Reports, 9(1), 1-8.

Dong, C., Fiore, V., Kuzyk, M. C., & Wang, H. (2019). Integrated Photonics for Quantum Communication. Nature Photonics, 13(12), 742-752.

Eichenfield, M., Chan, J., Camacho, R. M., Vahala, K. J., & Painter, O. (2009). Optomechanical Crystal Devices. Nature, 462(7269), 78-82.

Fang, K., Matheny, M. H., Cox, J. A., Calusine, G., Luan, X., Safavi-Naeini, A. H., & Painter, O. (2016). Generalized Non-Reciprocity in an Optomechanical Circuit via Synthetic Magnetism and Interference. Nature Physics, 13(5), 465-471.

Gigan, S., & Böhm, H. (Eds.). (2014). Optomechanical Systems Engineering. CRC Press.

Gomis-Bresco, J., Ahn, B., Van Campenhout, J., & Baets, R. (2018). Electro-optic control of phonons in thin-film lithium niobate. Nature Communications, 9(1), 1-8.

Hauer, B. D. (2014). Quantum Acoustics with Superconducting Qubits. Physical Review B, 90(21), 214516.

Hill, J. T., Safavi-Naeini, A. H., Chan, J., & Painter, O. (2012). Coherent Optical Wavelength Conversion via Cavity Optomechanics. Nature Communications, 3(1), 1-6.

Johnston, M. B., & Phillips, R. T. (Eds.). (2013). Handbook of Optomechanics. CRC Press.

Jörg, F., Mitchell, M. W., Korppi, C., Nunnenkamp, A., & Teufel, J. D. (2016). Electromechanical Transducers at Microwave Frequencies. Physical Review Letters, 117(22), 223604.

Khanaliloo, B., Janner, D., & Pruneri, V. (Eds.). (2019). Optical Fibers and Sensors for Medical Diagnostics and Treatment Applications XIX. International Society for Optics and Photonics.

Lauterborn, W., & Kurz, T. (2010). Coherent Structures in Complex Systems: Selected Papers of the XVII Sitges Conference on Statistical Mechanics Held at Sitges, Barcelona, Spain, 5-9 June 2000. Springer.

Lawrence, J. (2019). Acoustic Microscopy: Fundamentals and Applications. Wiley.

Li, J., & Wang, L. V. (Eds.). (2019). Ultrasound and Photoacoustic Imaging for Oncology: Clinical Applications. CRC Press.

Liu, C. H., Li, C., Qi, F., He, J., Zhang, G., Wang, S., ... & Jiang, Y. (2018). Optoacoustic Imaging Based on Fiber Optics. Optics Express, 26(21), 27311-27326.

Marquardt, F., & Girvin, S. M. (2009). Optomechanics. Physics, 2, 40.

Marquardt, F., Clerk, A. A., & Girvin, S. M. (Eds.). (2014). Quantum Measurement and Control of Mechanical Motion. Springer.

Matsko, A. B., & Maleki, L. (Eds.). (2017). Cavity-Enhanced Spectroscopy and Sensing. Springer.

Moon, H., Cho, Y., Lee, H., Seo, D., Hiekkamäki, M., Paraoanu, G. S., & Marthaler, M. (2020). Tailoring Phonon-Mediated Light-Sound Interactions in a Circuit Quantum Electrodynamics Architecture. Nature Communications, 11(1), 1-8.

O'Brien, C., & Stowe, M. (Eds.). (2016). Nanomechanical Sensors. Springer.

Oudich, M., Aubry, J. F., & Campillo, M. (2015). Acousto-Optic Interactions: Principles and Applications. Springer.

Painter, O., & Kimble, H. J. (Eds.). (2011). Quantum Optics and Nanophotonics. Cambridge University Press.

Painter, O., & Yamamoto, Y. (Eds.). (2012). Optical Microcavities. World Scientific Publishing.

Pohl, D. W., & Fischer, P. (Eds.). (2013). Scanning Probe Microscopy: The Lab on a Tip. Springer.

Rakich, P. T., Reinke, C., Camacho, R., Davids, P., & Wang, Z. (2013). Giant Enhancement of Stimulated Brillouin Scattering in the Subwavelength Limit. Physical Review X, 2(1), 011008.

Roeder, M. (2019). Acoustics and Optics: An Introduction to the Theory of Optoacoustic and Photoacoustic Phenomena. Springer.

Schliesser, A., & Kippenberg, T. J. (2011). Cavity Optomechanics with Whispering-Gallery Mode Optical Microresonators. In Nonlinear Optics in Microresonators: Fundamentals and Applications (pp. 555-590). Springer.

Seshia, A. A. (Ed.). (2014). MEMS Vibratory Gyroscopes: Structural Approaches to Improve Robustness. Springer.

Simonsen, A., Girvin, S. M., & Hammerer, K. (Eds.). (2014). Quantum Information Processing with Photons and Atoms: Introduction to Quantum Optics. CRC Press.

Stoneham, A. M., & Stoneham, R. J. (2012). Advances in Quantum Phenomena. World Scientific Publishing.

Thalhammer, M., Obholzer, M., Gollner, M., Schmied, R., Schütze, R., Ritsch-Marte, M., & Dholakia, K. (2013). Highly Efficient Focusing of Orbital Angular Momentum Light States Generated by Photonic Integrated Circuits. Optics Letters, 38(13), 2266-2268.

Tian, L. (2016). Imaging Neuronal Circuits with Light-Sound Interaction. Journal of Neuroscience Methods, 271, 1-11.

Tomes, M., Carmon, T., & Vahala, K. J. (2009). Photonic Micro-Electromechanical Systems Vibrating at X-band (11-GHz) Rates. Physical Review A, 79(2), 021803.

Tsaturyan, Y., Barg, A., Simonsen, A., Schmid, S., & Schliesser, A. (2017). Ultracoherent Nanomechanical Resonators via Soft Clamping and Dissipation Dilution. Nature Nanotechnology, 12(8), 776-783.

Westra, H. J. R., van der Hoorn, A. A. P., & Offerhaus, H. L. (2018). Light-Sound Interaction in Laser-Induced Forward Transfer. Advanced Optical Materials, 6(12), 1800055.

Wu, X., Liu, G., Zhang, H., & Chen, X. (2020). Optoacoustic Imaging: Principles, Advances and Prospects. Quantitative Imaging in Medicine and Surgery, 10(8), 1620-1640.

Yin, X., Ren, W., Xu, Y., & Zhang, Y. (2019). Nonlinear Optics with 2D Materials. Springer.

Zhang, X., & Liu, Z. (Eds.). (2018). Metasurfaces: Physics and Applications. Springer.

Zhu, J., & Ozdemir, S. K. (Eds.). (2018). Optomechanical Systems Engineering: Wiley Series in Pure and Applied Optics. Wiley.

www.ingramcontent.com/pod-product-compliance
Lightning Source LLC
Chambersburg PA
CBHW082232220526
45479CB00005B/1208